Chemical Reaction Engineering—Plenary Lectures

Chemical Reaction Engineering—Plenary Lectures

James Wei, Editor
Massachusetts Institute of Technology

Christos Georgakis, Editor
Massachusetts Institute of Technology

Based on the 7th
International Symposium
on Chemical Reaction Engineering
in Boston, Massachusetts,
October 4–6, 1982

ACS SYMPOSIUM SERIES 226

AMERICAN CHEMICAL SOCIETY
WASHINGTON, D.C. 1983

Library of Congress Cataloging in Publication Data

International Symposium on Chemical Reaction Engineering
(7th: 1982: Boston, Mass.)
Chemical reaction engineering—plenary lectures.

(ACS symposium series, ISSN 0097-6156; 226)

Includes bibliographies and index.

1. Chemical engineering—Congresses. 2. Chemical reactions—Congresses. I. Wei, James, 1930– . II. Georgakis, Christos, 1947– . III. Series.

TP5.I67 1982a 660.2'99 83-11876
ISBN 0-8412-0793-3

PRINTED IN THE UNITED STATES OF AMERICA

ACS Symposium Series

M. Joan Comstock, *Series Editor*

FOREWORD

The ACS Symposium Series was founded in 1974 to provide a medium for publishing symposia quickly in book form. The format of the Series parallels that of the continuing Advances in Chemistry Series except that in order to save time the papers are not typeset but are reproduced as they are submitted by the authors in camera-ready form. Papers are reviewed under the supervision of the Editors with the assistance of the Series Advisory Board and are selected to maintain the integrity of the symposia; however, verbatim reproductions of previously published papers are not accepted. Both reviews and reports of research are acceptable since symposia may embrace both types of presentation.

CONTENTS

PREFACE

IN THE NARROW SENSE, CHEMICAL REACTION ENGINEERING (CRE) is concerned with the design, operation, optimization, and control of processing equipment called chemical reactors. CRE is the counterpart to unit operations (UO), which is concerned with processing equipment in which no chemical reactions take place. Both chemical reaction engineering and unit operations integrate many disciplines to attain practical engineering goals. They differ from the engineering science subjects of kinetics, thermodynamics, fluid mechanics, and transport phenomena that seek to isolate and analyze an individual aspect of a complex piece of machinery or to understand and describe quantitatively the material behavior at microscopic or molecular scales.

The main difference between chemical reaction engineering and unit operations lies in the presence or absence of chemical reactions. Except for purely separational processes, such as the production of oxygen from air, the heart of any chemical process is one or more chemical reactions that transform less valuable feed material to more valuable products. In the case of combustion, the main products are heat and energy rather than material products. In the case of pollution control equipment, the conversion of a harmful feed to harmless products is the main objective. Chemical reactors lie in the heart of any chemical plant, and the unit operations equipment serves the needs or compensates for the inadequacies of the chemical reactors. The design of a chemical plant must start with the chemical reactors, which usually run 10–20% of the total plant cost. When you look over the flowchart of a process, you will find the chemical reactor as a small unit in a very large network; yet, the design of all other equipment revolves around the chemical reactors. The optimum design of a chemical reactor does not necessarily lead to the optimum design of a chemical plant. The peripheral equipment of unit operations (the tail) usually wags the chemical reactors (the dog). A good design engineer chooses the dog that has the smallest tail.

In a comparison of the salient features, the following may be noted:

	Chemical Reactors	**Unit Operations**
Function	Chemical reactions	Physical operations needed to service the reactors
Design	Individually tailored for each reaction	Modular; often can be ordered from catalogs
Cost	10–20% of total plant	80–90% of total plant
Performance	Highly nonlinear due to kinetic dependence on temperature and concentrations	Nearly linear
Intellectual Collaborators	Chemists and chemical engineers	Mechanical engineers

In the broader sense, chemical reaction engineering is also concerned with the components needed for the synthesis of chemical reactors, such as kinetics, catalysis, catalyst design, mixing, multiphase systems, fluid mechanics and transport in fixed and fluidized beds, and solid movement mechanics in moving and fluidized beds. Study of the maldistribution of gas and liquid flow in trickle beds can be considered as a branch of fluid mechanics, but this subject is very seldom studied by the major researchers in fluid mechanics. However, it is a subject that is studied by major researchers in reaction engineering. On the other hand, paper-making machines would seem to be very suitable subjects for reaction engineers, but are hardly ever studied. Polymerization reactors are enormously important in the chemical industries and are beginning to attract serious researchers with backgrounds in reaction engineering.

Reaction engineers form a living and changing tribe that keeps in touch through professional meetings and journals. They share a common set of textbooks and tools, the same vocabulary and analytical scheme to analyze problems, and the same set of triumphant classic achievements in the past. They agree on what are proper problems to be tackled and what are proper standards for admissible reports and papers. In short, they form a paradigm of their own. New people are constantly recruited by this tribe through the process of apprenticeship and junior partnership, or by adoption into this tribe of brilliant outsiders with brand new ideas. Whatever is accomplished by a member of this tribe will be shared with the rest of the members and is by definition chemical reaction engineering.

In this volume, we print the six plenary speeches given at the 7th International Symposium on Chemical Reaction Engineering at Boston. First, the intellectual foundation of chemical reaction engineering is described by Aris, and second, the study of alternative reactors for the methanol-to-gasoline process is described by Penick. These two chapters form a powerful contrast between the academic intellectual approach and the industrial pragmatic approach. Fluidized beds are reviewed by Rowe, and fluidized-bed combustors are reviewed by Sarofim. These two chapters are concerned with one of the most powerful reactors, in which scale up by design engineers is still done with great trepidation, and one of its most important future applications. If fluidized-bed combustors for coal are well engineered, there will be little demand for clean synthetic fuels, except for liquid fuels for transportation. Combustion should have been one of the most central problems in reaction engineering. Only historical events decree that it should be a meeting ground of chemical engineers, mechanical engineers, and chemists with their own Combustion Institute and highly successful International Symposium series. It is overdue for us to build bridges to these outstanding people. Micromixing is reviewed by Villermaux, and polymerization reactors are reviewed by Ray. These two chapters are concerned with homogeneous phase and nearly homogeneous phase reactors that are of growing importance. Most textbooks of reaction engineering have rather short sections on polymerization, a subject found more often in polymer textbooks. These reviews will go a long way toward remedying this defect.

The plenaries serve as a powerful centralizing force among reaction engineers, who are always in danger of moving off in their own individual rivulets and drying out. These plenaries should be read by all students of chemical reaction engineering, whatever their subspecialty. They augment the basic textbooks as part of "what every educated reaction engineer should know."

JAMES WEI
Massachusetts Institute of Technology
Cambridge, MA

October 22, 1982

1

Chemical Reaction Engineering as an Intellectual Discipline

R. ARIS

University of Minnesota, Department of Chemical Engineering and Materials Science, Minneapolis, MN 55455

What, you are entitled to ask, do you mean by an intellectual discipline? Before attempting any formal answer, let me turn to two of the traditional disciplines and try to sketch some of the characteristics that they may have or induce in their adepts.

I take my text from the advice given to President Gilman, the first of the Johns Hopkins University, to start with the best classical scholar and the best mathematician that he could find. The latter was none other than the great James Joseph Sylvester, retired from Woolwich since 1870 and spending his time in the enjoyment of the classics, playing chess and versification on the principles of his "Laws of Verse"—a pamphlet by which he set great store. With his appointment to Hopkins at the age of 62 there came the second flowering of his genius and with it his exploration of the fundamental system of invariants and the syzygies of algebraic forms.

Sylvester's career had not been an easy one (1). After his studies at Cambridge had been interrupted by illness he took his degree in 1837 as Second Wrangler in the same class as George Green. To say that he took his degree is not quite accurate, for Sylvester, who says of himself that he was one of the first holding "the faith in which the founder of Christianity was educated" to compete for the mathematical tripos, could not complete his degree without subscribing to the 39 Articles of the Church of England—a subscription he was unwilling to make. He therefore went to Trinity College, Dublin from which he received his degrees in 1841. His Cambridge degree he did not receive until 1872, when the religious barriers had at length been removed. In 1890 he was given an honorary Sc.D. at the same time as Benjamin Jowett, Henry Perry Liddon and other notables. The Public Orator could, by then, bracket him with Newton as "Sylvester noster" in the accolade: "Nonnulla quae Newtonus noster, quae Fresnelius, Iacobius, Sturmius, alii, imperfecta reliquerunt, Sylvester noster, aut elegantius explicavit aut argumentis veris comprobavit." After two years at University College, he crossed the Atlantic to the University of Virginia but

0097-6156/83/0226-0001$06.00/0

remained there only a matter of months (2). Returning to England he spent some time as an actuary for a life insurance company, a drudgery which might have stifled the mathematical genius of a less robust man. During this period, however, his friendship with Cayley matured and he took private pupils, the most illustrious of whom was Florence Nightingale who went out to Crimea in 1854, the year of Sylvester's release from the legal world. He returned to academic life as professor of mathematics at Woolwich with a salary of £550, government quarters and the right of a pasturage on Woolwich Common. He continued there until 1870 when he retired over what he regarded as an unfair change in the regulations and with some bitterness over his pension. Perhaps it was this that made him, in coming to Johns Hopkins, request that his salary, the considerable sum of $5,000, be paid in gold (3). At all events the move for the second time to America was a great success and he found himself able to teach mathematics with great freedom and play a leading role in the intellectual life of Baltimore society.

His colleague in classics was also of the greatest distinction. Called from the same university in which Sylvester had had so unhappy an experience years before, Basil Lanneau Gildersleeve was a Southerner born and bred who served with the Confederate Army while the Civil War was in progress during the summer months in order (as he says) "to earn the right to teach Southern youth for nine months...by sharing the fortunes of their fathers and brothers at the front for three." Much later in the Atlantic Monthly of January 1892 he still was "not certain that all" readers might "appreciate the entire clearness of conscience with which we of the South went into the war" (4). I have not been able to find any reference to the direct intercourse of these two men, though I would think that, with the exception of the question of states' rights, they would have had much in common. It was Sylvester's delight to read the classics and Gildersleeve would have approved the motto which he gave the American Journal of Mathematics which was founded at Hopkins in 1878 and edited by Sylvester:

πραγμάτων ἔλεγχος οὐ βλεπομένων

(the evidence of things not seen)

Gildersleeve does refer to Sylvester's metrical theory of "phonetic syzygy" in one of the Brief Mentions of the American Journal of Philology, which he founded in 1880 early in his time at Hopkins and published from there for many years—a pleasant foil to Sylvester's American Journal of Mathematics. Gildersleeve was, of course, a much younger man—45 to Sylvester's 62—when he came to Hopkins but, in contrast to Sylvester's brief tenure of 8 years, he presided over the destiny of the classics there for the next 39. During that time he established graduate studies in the classical disciplines, directing no less than 67 doctoral dissertations thus having a profound influence on classical studies throughout the country. In 1901 some forty-five of his old pupils put together a volume of largely philological studies in his honor (5). Like Sylvester, he was a prolific publisher on a wide range of subjects. Sylvester's "Collected Works" run to 4 quarto volumes, edited by H. F. Baker in 1912; Gildersleeve's papers were as numerous though not as completely collected. A volume of his studies and essays appeared in 1890 (6) and a selection of his Brief Mentions published over the years in the American Journal of Philology was edited with a biographical sketch by his successor and devoted pupil in the chair of Greek at Johns Hopkins (7). Gildersleeve published more books than Sylvester and his Latin grammar, revised with the cooperation of Professor Lodge, is still a standard reference book.

Both men rejoiced in an amplitude of style that betrays a more leisured age than ours. For example, in adverting to the then recently published book entitled "Value of the Classics" (Princeton University Press), Gildersleeve remarks that "appended to these impressive deliverances there is a formidable array of statistics drawn up in refutation of those other statistics that have been used only too effectively to stir up the pure minds of believers in Latin and Greek. My missionary days are long overpast and statistics have lost whatever charm they had for me even in the syntactical line, but the book will strengthen the feeble knees of those who are afraid that they will have to bow themselves down in the House of Rimmon." Compare Sylvester. He published a paper in the American Journal of Mathematics in 1882 entitled "A Constructive Theory of Partitions, Arranged in Three Acts, and Interact and an Exodion." Act One is on partitions regarded as entities and is prefaced with the quotation from Twelfth-Night..."seeming parted, but yet a union in partition." In a paper on the three laws of motion in the world of universal algebra published in 1884 he has a charming footnote in which, after referring to a theorem in the text, he remarks "I have not had leisure of mind, being much occupied in preparing for my departure, to reduce this theorem to apodictic certainty. I state it therefore with due reserve."

But we must turn from the biographical details of these giants, however fascinating they may be, and ask about the characters of their subjects as disciplines. The very name "classics" is an acknowledgement of our debt to the cultures of ancient Greece and Rome, for in using it we are reminding ourselves that it is they who first recognized the classes, or categories, of thought that we, for all our neglect of their sources, still observe today. Whether we like it or not, the furniture, which we bump into (to borrow a metaphor from J. K. Newman of Illinois) in our gropings through the room of intellect, was built and put there by the Greeks. In our everyday and scientific language we pay hidden tribute to the notions of classical times using words that were consciously coined to reflect their thought. Thus "gas" for the third estate of matter has now been common for more than three centuries but goes back to the alchemical notion of the indwelling spirit of a thing, which

Van Helmont (1577-1644) in his "Ortus Medecinae" (1652) referred to as "halitum illum Gas vocavi, non longe a Chao veterum secretum." This was charmingly translated in Chandler's "Van Helmont's Orations" (1662) as: "I have called that vapour, Gas, being not far severed from the Chaos of the Auntients...a far more subtile or fine thing than vapour, mist or distilled Oylinesses, although as yet it be many times thicker than air. But Gas itself, materially taken, is water as yet masked with the Ferment of composed Bodies."

If 'gas' in the physical sense still bears echoes of the formless darkness over the face of the deep, the currently popular study of mathematical chaos uses the Greek work in simple transliteration. Yet it is pleasant to reflect that a recent meeting on this subject in Los Alamos (May 1982) could be subtitled "ΚΟΣΜΟΣ ΕΝ ΧΑΩ" and thus make ironic comment on the world as well as recalling the original sense of the beauty of order inherent in the word κόσμος. For of its original meaning—adorning or ornament—the only trace left to us in English is a superficial, almost meretricious, one—cosmetic. But that it should also mean order, a 'lucidus ordo', is natural enough and was so transferred by Pythagoras to the world as expressing that ordered beauty whose appreciation is the basis for all scientific—not to say humanistic—endeavour (8). 'Mundus' in Latin follows a parallel course as we see in Pliny's comment: "Quem κόσμον Graeci nomine ornamenti appellaverunt, eum nos a perfecta absolutaque elegantia mundum": or Cicero's: "Hunc hac varietate distinctum bene Graeci κόσμον, nos lucentum mundum nominamus".

To study of the classics is surely to submit to a discipline that rejoices in the relation of words and ideas. It is not just to luxuriate in etymological overtones (a vice into which I have just allowed myself to be betrayed) but to learn the relation of word and thought as expressed by the precise usage of words and their place in the syntactical structure of a sentence. The act of translation from one language into another demands this hard striving for precision, and is particularly called forth by an inflected tongue and one, such as Latin, where the thought is marshalled into principal and subordinate clauses. By contrast English, for all its merits, has a lax and individual structure, rambling comfortably on in coordinated measures. It is almost like the contrast you can experience by turning aside from Watling Street or the Fosse Way into a country road or lane as you go "to Birmingham by way of Beachy Head." The very familiarity of our native speech and its colloquial use—correct and effective though this be—denies it the cutting edge needed to shape our minds. As Livingstone wrote years ago "No doubt there are more important things in education than the study of grammar; but it is not an overstatement to say that not to know Greek is to be ignorant of the most flexible and subtle instrument of expression which the human mind has devised, and not to know Latin is to have missed an admirable training in precise and logical thought" (11).

Of the converse discipline, that of turning English into Latin or Greek it is interesting to quote the Physics master at Winchester. "The task", he says, "involves concentration, close attention to detail, and considerable logical reasoning; there are no short cuts, no formulae as in the science problem, the reasoning involved cannot be avoided by mere effort of memory as in the writing out of a proposition in geometry" (12). It has been popular among great men from Winston Churchill to Peter Danckwerts to deplore the hours spent on Latin and extol their instruction in English, but, as I have ventured to comment elsewhere in reviewing "Insights into Chemical Engineering" (13), their English may owe more to the discipline of the classics than they affect to admit.

But the study of the classics is not merely a matter of language however great the general benefit to our general understanding of language this may afford. Language is the entry to literature as literature is the reflection of the life and thought of its time. From their place at the root of western civilization, the Greek and Latin cultures have an uncanny way of anticipating and illustrating modern political and social problems. Classic indeed is Thucydides' description of the internal corrosiveness of war and revolution "Words had to change their ordinary meaning and to take that which was now given them...Frantic violence became the attribute of manliness; cautious plotting a justifiable means of self-defense. The advocate of extreme measures was always trustworthy; his opponent a man to be suspected" (14). As sociology or social psychology, its penetration should be the envy of the modern social scientist. As literature, its quick antitheses "breathe an almost human aliveness" into an otherwise abstract analysis (15). Archaeology reveals the circumstantial life as literature does the intellectual, and it too has its own disciplines—its sherds often to be handled as delicately as particles, its artifacts as lovely as odes and epodes, and its sites as complex and intricate as the plot of a tragedy or interplay of a dialogue.

This, however, is not the place to extol the classics—nor do they need it—what I am concerned with is the features of intellectual life that their study cultivates. In the first place, there is the philological foundation. This necessary occupation with words--their precision of meaning & the exactness of the way they are put together--is fundamental to the careful reconstruction of a text and its critical evaluation. A training in this sometimes leads to a certain cautiousness and care in quotation. Housman gave his inaugural as Kennedy Professor of Latin at the University of Cambridge on 9 May 1911, but he never published it as was customary, for he was never able to 'verify a statement it contained as to the text of Shelley's Lament of 1821' (16). The manuscript was destroyed after his death but a further copy found among his papers was published in T.L.S. in 1968, omitting, however, the passage of Shelley (and Swinburne's remarks on it) that he had been unable to verify (17). It was not until after further work on Shelley manuscripts had justified the quotation as Housman had used it, that John Carter published the lecture as "The Confines of Criticism" in 1969 (18). I mention this example not merely because of this care for exactness, which by some might be thought excessive, but because Housman used it to show the kind of confusion that can arise if exactitude of this order is not maintained. Swinburne had praised an unmetrical line in the commonly printed version of the second stanza as "a thing to thrill the veins and draw tears to the eyes of all men whose ears were not closed against all harmony" by "the melodious effect of its exquisite inequality." But the line that so thrilled Swinburne's veins was a misprint and the "sovereign sweetness" of the stanza not Shelley's craft but some unknown compositor's carelessness. "These, "added Housman severely," are the performances of the literary mind when, with its facile emotions and its incapacity for self-examination, it invades the province of science".

It is only fair to add that having smitten the English too-literary carelessness he went on to denounce the German methodisch unimaginativeness; the one breeding the fatuousness of "taste", the other the mentality of the slave. Against neither could he give an infallible alexikakon, though he warned against that "servility shown towards the living" that "is...so often found in company with lack of due veneration towards the dead" (18, p. 44).

If the classics demand exactness they do not rule out imagination—the textual critic, literary critic, philosopher and archaeologist all need it—but the discipline demands that this imagination should be controlled by precision of scholarship rather than the ramification of fancy. To the nice appreciation of language the classical scholar must add a certain copiousness of learning. A very broad reading of literature, history and philosophy is part of the graith of classical scholarship just as a wide experience of different aspects of ancient civilization and a keen and discerning eye for its artifacts is needed by the archaeologist. Indeed the contrast that most forcibly strikes a scientist in talking to the humanist is the comparatively immense amount of reading that the latter must do and the recourse that must be had to primary sources. Sir Andrew Aguecheek's lament in "Twelfth Night" (provoked you will recall by Sir Toby's "Pourquoi") "O that I had but followed the Arts" (19) is a real cry of the heart, for, save for the exceptionally gifted, the acquisition of the varied and extensive knowledge which is needed in ancient studies is an impossible task if not begun in childhood.

But let us turn to a discipline which, as Hardy says, is "more than any other art or science,...a young man's game"—mathematics (20). As an intellectual discipline, its leading characteristics are the power of abstraction, the passion for elegance and the pursuit of rigour. I put them in that order, for, though they share these characteristics in some degree with other disciplines, the power of abstraction in mathematics is raised to a height unattainable in any other subject, save perhaps music. As a result connections can be found between things which are very far from obvious. It was, as Whitehead pointed out (21), an immensely important step forward in the history of thought when it was first realized that there was a commonality of some sort between a pair of twins and a brace of pheasants. We take the concept of number so completely for granted that we tend to forget what an achievement this abstraction of concept must have been. Perhaps we can recall the initial confusion and ultimate delight of recognizing that the cardinal number of a class was indeed the class of all classes similar to it, or at a more lowly level our liberation from x and y as apples and oranges in our kindergarten struggles with elementary algebra.

It is of course a long way from the five and country senses of the engineering world to the thin air of the heights that mathematicians seem to breathe so effortlessly, yet we are fortunate to have had as part of our technical training at least a brush with some of the structures that pervade mathematics. The group, ring or field in algebra with the notions of equivalence relations, morphisms, and mappings are part of our mental furniture, not only in their instantiations, but as abstract entities forming our powers of apprehension. It is thus that we perceive that, from a purely stoicheiometric point of view, any non-singular transformation of a set of chemical reactions gives an equivalent set. At the same time we are conscious of the fact that stoicheiometry is not everything and that we may have to move beyond it to an aspect in which a particular set has a preferred character. Not to have perceived the equivalence is to be left in a confusion of concretions; not to have perceived its limitations is to have failed to understand the nature of abstraction.

I suppose the root of these notions goes back to Pythagoras who discerned the pervasive influence of number in the order of nature. Of course this was not developed until much later when, with the understanding of the notion of function, it became possible to express physical laws in mathematical form, but his proof (or rather the proof which goes back to his school in the sixth century B.C.) of the irrationality of the square root of two is cited by G. H. Hardy (20) as opening up the whole realm of real numbers, making that definitive step beyond the everyday commonplaces of integers and rationals.

If I may, I would like to advert for a moment to the recent development of non-standard analysis and sketch how infinite and infinitesimal numbers can be presented. In this I follow a beautiful expository article of Ingleton (22) though, in my haste scarcely doing him, or Luxemburg on whom he leans, full justice. Consider all infinite sequences of real numbers X $(x_1, x_2, \cdots, x_n, \cdots)$ and let two such entities be equivalent if they differ only in a finite number of elements i.e., $X \equiv X'$ if $x_n = x'_n$ for all but a finite number of n. From now on we can consider the entity X to be the equivalence class and representable by any of its members just as the rational 1/2 is the class (1/2, 2/4, 3/6,..). We retain the real numbers x as the elements $(x,x,\cdots,x\cdots)$(i.e. $x_n = x$ for all n) and can define arithmetical operations such as $X \pm Y = (x_1 \pm y_1,\cdots,x_n \pm y_n,\cdots)$ $X.Y = (x_1y_1,\cdots,x_ny_n,\cdots)$ and $1/X = (1/x_1,\cdots,1/x_n,\cdots)$ provided $x_n \neq 0$—if it is, ignore it and use 1 instead of $1/x_n$ since this can only happen a finite number of times. These definitions allow us to comprise the ordinary operations of the arithmetic of reals within this new world of hyperreals. If we use the usual notation for the reals we can say that $x \varepsilon R$, $x \equiv (x,x,\cdots,x,\cdots)$ provides a mapping between R* the space of hyperreals $X = (x_1,\cdots,x_n,\cdots)$ and the rule $f(X) = (f(x_1),\cdots f(x_n)\cdots)$ is a natural extension of the function f from R to R*.

But now we notice that we have in R* a much richer entity than we had in R. We have hyperreal numbers like $Z = (1,2,3,\cdots,n,\cdots)$ which is in any sense of the word an infinite integer. Yet we have not the undifferentiated world of aleph-zero, in which, you recall, $\aleph_0 = N\ \aleph_0$ or $\aleph_0 = N + \aleph_0$ for any finite N. But here $Z+N = (N+1,\cdots,N+n,\cdots)$ and or $N.Z = (N,2N,\cdots,nN\cdots)$ are well-defined and give $(Z+N) - N = Z$, $(Z+N) - Z = N$, $(N.Z)/N=Z$ and $(N.Z)/Z = N$. Similarly $Z^2 = (1,4,\cdots,n^2)$ or $Z^Z = (1,4,27,\cdots,n^n,\cdots)$ are truly "infinite" numbers. On the other hand $1/Z = (1,1/2,\cdots,1/n,\cdots)$ is an "infinitesimal", as are $1/(Z+N)$, $1/Z^2$ etc. Sums of infinitesimals are infinitesimals, as are products of infinitesimals with finite hyperreals. Products of infinite and infinitesimal numbers can be finite however as can be quotients of infinites or of infinitesimals. Two hyperreal numbers are infinitesimally close if their difference is infinitesimal. For example $(Z+1)/Z$ and $Z/(Z+1)$ are both infinitesimally close to 1 though all three are <u>distinct</u> hyperreal numbers. In fact every finite hyperreal number is infinitesimally close to a real number and every real number has an infinity of infinitesimally close hyperreals. The real number infinitesimally close to any hyperreal is called its standard part, i.e. $stX = x$ if $(X-x)$ is infinitesimal.

This makes the discussion of convergence rather simple. A sequence y_n in R is a mapping from N the space of integers to R. This can be extended in the non-standard reals as a mapping from the extension of N to N*, the hyperreal integers, into R*. Then Y_M is an element of a non-standard sequence if $Y \varepsilon R^*$ and $M \varepsilon N^*$. We say that Y_M converges to Y if $Y = st\ Y_M$ for all infinite integers M. Continuity and differentiability can be similarly defined, for example $f'(a) = st\ \{f(a+h) - f(a)\}/h$ for all infinitesimals h. The Dirac delta function can be defined pointwise and new approaches to the theory of probability and stochastic differential equations are opened up. But these are applications and paradoxically I have been describing the abstraction that is the non-standard world in very concrete terms. This is no place to go back and try to present it abstractly, but it stands as an example of a recent step forward in the path of abstraction that began two and a half millenia ago (see also (<u>23</u>,<u>24</u>)).

In describing the characteristics of mathematics I put elegance before rigour, for so it appears to the mathematically cultivated mind. It is not that sloppiness of reasoning can be tolerated in the least, but rather that rigour is a discipline learned with care and practiced with scrupulousness, whereas it is the imaginative leap that reaches out to a result and feels that it is right before the perhaps lengthy confirmation is undertaken that the mathematician chiefly prizes. In this leap the driving force is imaginative rather than logical and it is the instinct for beauty which is the avenue to truth. Thus elegance—a much deeper quality than prettiness, though more restricted by context than beauty in the larger sense—is greatly valued in mathematical circles. In its highest forms, mathematics has "the simplicity and inevitableness" of great art and, vastly though he admired him, Hardy's judgment, as a mathematician, of the mathematician Ramanujan was that his work fell short of this—"it would be greater if it were less strange" (25) was Hardy's way of expressing it. Yet he paid full tribute to Ramanujan's "profound and invincible originality".

Though denying that mathematicians have consciously aimed at beauty rather than logic, Russell (26) has written eloquently of the aesthetic appeal of mathematics in one of his early essays.

> "Mathematics, rightly viewed, possesses not only truth, but supreme beauty—a beauty cold and austere, like that of sculpture, without appeal to any part of our weaker nature, without the gorgeous trappings of painting or music, yet sublimely pure, and capable of a stern perfection such as only the greatest art can show. The true spirit of delight, the exaltation, the sense of being more than man, which is the touchstone of the highest excellence, is to be found in mathematics as surely as in poetry. What is best in mathematics deserves not merely to be learnt as a task, but to be assimilated as part of daily thought, and brought again and again before the mind with ever-renewed encouragement."

But from the realm of the classics, so rich and warm with human associations, and the "cold and austere" beauty of mathematics we must turn to the banausic domain of chemical reaction engineering. It was, if I remember rightly, at one of these meetings that Professor Wei (27) gave us the connection by pointing out that one of the oldest technological processes is the fermentation reaction, as pure a reactive process as we could wish for—rich and humane whilst it stayed within our domain and took not up with distillation processes—and rightly respected in the very title of our meeting. But important though the business of wine making may be it is but one of many reactions which lie at the cores of the different processes of chemical industry and it is out of the experience of these that the discipline of chemical reaction engineering has grown. It is a recent growth in a subject of no long history for I suppose that it is only in the last fifty years that it has emerged as a distinguishable train of thought. Damköhler's work in the 30's, Hougen's and Watson's in the 40's, Wilhelm's and Amundson's in the 50's are but points of touch-down in the immense strides that have taken us onward. The tale of texts has lengthened decade by decade, the wen of papers has grown year by year and I hesitate to mention any more names for I am sure to leave out people of great significance. (I have tried to survey some aspects of the historical development elsewhere (28).)

But what has emerged as characteristic in this development? I would suggest first a sense of balance, then a feeling for structure and finally a refinement of concept leading to a high degree of awareness. Those are rather vague terms so let my try to make them more precise and then illustrate some aspects in greater detail. Under balance I think of the balance that needs to be struck between theory and practice on the one hand and between the general and the particular on the other. By structure I mean the way in which we have come to see the subject standing on the foundation of thermodynamics, transport processes and chemical kinetics. The nature of the constraints implied by this and the inherent structure of, for example, stoicheiometry or mass action kinetics are at issue here. Certain ideas have been of great importance in this development and no part of chemical engineering is richer in dimensionless numbers; they range from the four of Damköhler (29) or the modulus of Thiele (30) to the Carberry number that Bischoff (31) minted a year or two ago. It was early recognized that reactors presented peculiar scale-up problems and the height of reaction unit never attained the popularity of the H.T.U. Simplified structures play an important role as in the insight gained from Wei and Prater's (32) monomolecular systems and the resulting shape invariance (33). A great refinement of concepts has taken place over the years as, for example, in the way that we now distinguish sensitivity, multiplicity and stability (34). With the exploration of instances of these concepts has come an awareness of the varieties of behavior, the comprehensive description of all forms of possible behavior of reactors as dynamical systems the pervasiveness of oscillation and chaos and, with all this, an exhanced ability to feel after the form of the solution.

Chemical reaction engineering has been notable for the balance between experiment and theory that has marked it from the beginning. It is true that, save in the rare instance of a Shinnar at the industrial-academic interface or a Schmitz in the more purely academic context, this blood and judgment have seldom been "so well commingled" in one person, but there has always been a propitious intercourse between those chiefly concerned with practice and those pursuing the euristics of mathematical modelling. The ingenuity of the practicing reactionary is always refreshing: on the industrial scale we have the classical case of the fluidized bed as well as the serendipity of the transfer line or the intricacy of some of the polymerization reactors; on the research scale there is Wicke's creeping zone (35), Rony's bundle of microfilaments (36), Petersen's single pellet reactor (37), Carberry's spinning basket (38) and the Berty reactor, the string of beads used by Hegedus and Oh (39) and Wei and Degnan's (40) combination of cross-flow monoliths or Schmitz' polythene bag. It is a practical ingenuity matched on the theoretical side by some of the devices that have been used in analysing reactor models. The notion of an effectiveness factor introduced by Thiele, Amundson's exploitation of the phase plane (34), Gavalas' use of the index theorem (41), the Steiner symmetrization principle used by Amundson and Luss (42) and the latter's exploitation of the formula for Gaussian quadrature (43)—perhaps the prettiest connection ever made in the chemical engineering literature—are theoretical counterparts, large and small, of the careful craft of the experimentalist. So perhaps also the very important insight that Danckwerts contributed in his formulation of the residence time distribution is a happy foil to his heroic ambition to trace a blast furnace (44).

Faced with the enormous variety of practical reactor, chemical engineers are forced to think of the general shape and structure of reactor analysis if they are to retain it as an intellectual discipline. Wilhelm used to speak of the "morphology" of the subject years ago in his lectures at Princeton and his influence on this side of the Atlantic, together with that of Hougen and Watson, was as formative on the engineering side, as Amundson's was on the theoretical (45). There are many levels of structure from the rather obvious fact that stoicheiometry is a manifestation of linear algebra to the profound analysis of mass action kinetics that Feinberg (46) has developed so fully from his earlier work with Horn and Jackson. Wei and Prater's (32) now classic treatment of first-order reactions showed how the structure of linear differential equations leads to both theoretical and practical insights into kinetic systems. More recently Shinnar has shown how fundamental thermodynamic constraints, such as the requirement that the free energy decrease along a reaction path, give shape to the problem of understanding a reactive system and incorporating it into useful design objectives.

The introduction of singularity theory by Golubitsky and Keyfitz (47) and its development by Luss and Balakotaiah (43,48) is another structural landmark. The latter show, for example, that all the possible bifurcation diagrams can be determined for certain networks of reactions. They use a reductive scheme which allows the system with N reactions to be analysed by limiting cases in which only n reactions proceed at a finite rate and the others are either instantaneous or glacially slow. It would be inappropriate, even if it were necessarily relevant, to say more on this since a paper at this meeting will bring us up to date on their thinking. But it raises two points of importance. The first is the instinct for comprehensiveness; the second, the immensity of many dimensions, which I will advert to at the end. As a practical device for making a product, the chemical reactor presents itself as a matter of design and control and the engineer may be satisfied to obtain economy in the one and reliability in the other. It is not a question of considering all possible reactors but of achieving a working reactor that will be integrated into a whole plant and come on stream in time to make a profit. As an object for intellectual comprehension, however, the reactor must be regarded comprehensively, we have the right and duty to enquire of its behavior for any, not just the so-called "realistic", values of the parameters. This is where such tools as singularity theory or that of catastrophes, the qualitative theory of differential equations or topological methods are so important, for they can, when rightly used, often give comprehensive information. A case in which the strict numerical definition of regions of parameter space could be replaced by geometrically distinct possibilities arose with the competition of two microbiological populations for a single nutrient in a continuous fermentor which Arthur Humphrey and I considered a few years ago (49). We were able to show how the relative dispositions of the two growth curves and the point representing the dilution rate and feed concentration could be encoded into a label which was associated with a phase portrait. Thus it could be said that the dynamical behavior of any such system was topologically equivalent to any other with the same label.

It is tempting to digress at this point and enlarge on the virtues of chemical reaction engineering as a source of material for the craft of feeling out a solution. This is as important a technique to learn and teach as it is enjoyable in its exercise. However it would take us too far afield and, as I have tried to expound it elsewhere (see Chap. 2 of 28; 50,51), I will resist the temptation even though I have a brand new example making propitious ferment in my vitals.

A further characteristic of chemical reaction engineering is its facility in refining concepts. It is a derived characteristic, stemming from the mathematical component of the subject, but it has been typical of its development since the distinction between stability and sensitivity was first drawn (34). There have been lapses, alas, as in the case of an expository paper (52) which, to judge by the demand for offprints, proved quite useful, but which rated a very low Watkins number (53), for its title was "On the stability criteria of chemical reaction engineering" while its burden was as much the multiplicity criteria as the stability conditions. But there has been a steady incorporation of precisely defined concepts as for example in the burgeoning study of chaos which is evoking interest from mathematicians, physicists and chemists as well as engineers. In the 50's it seemed likely that oscillations in reactors would be most likely to arise when a control system was added to the reactor (54). The notion of multiplicity carried with it the picture of the associated region of attraction, an open set of initial states which would lead to the same steady state. Denbigh (55) had shown this in his pioneering work using the notion of "equifinality" adumbrated in a biological context by Burton. But, using the newly revived notions of Hopf bifurcation, Uppal, Ray and Poore (56) were able to show that the uncontrolled non-adiabatic stirred tank with a single exothermic reaction—the simplest non-isothermal system—could exhibit a wide variety of behavior including multiple steady states and limit cycles within limit cycles. Their tale of possible modes of behavior was scarcely told before a group of Russian authors (57) augmented it.

An important realization that has been attained in recent years is that the processes of adsorption, desorption and rearrangement on the catalyst surface may themselves produce multiple reaction rates or oscillations. Wicke (35) and his colleagues found this some years ago and the reviews of Sheintuch and Schmitz (58) and Scheintuch (59) show how wide-spread the phenomenon is. It is dangerous to mention names when I am sure to omit many important ones but those of Yablonskii, Slin'ko and their colleagues in Russia, Eigenberger and Hugo in Germany, Kenney in England and Luss, Takoudis, Schmidt, Ray, Jensen over here spring to mind in addition to those already mentioned. The Belousov-Zhabotinskii reaction has of course generated a minor industry of experimental and theoretical studies, dissipative systems and locus-ators which it would be rash to rush into here.

But even as the notion of periodic solutions to the nonlinear differential equations of chemical reactor theory was growing increasingly familiar in the '60's and '70's, there was sprouting beside it the new ideas of chaos and non-periodic behavior. This stemmed from a vastly simplified meteorological model of three modestly nonlinear equations studied by Lorenz (60) which exhibited solutions which were neither periodic nor asymptotic to periodic solutions. The properties of such solutions soon attracted attention of mathematicians and others and there is a thriving industry trying to put order into chaos at the moment (61). Already at the last of these symposia, Pismen (62) in his excellent review of kinetic instabilities was able to give several references and has himself shown how data sampling can contribute its meed of randomness.

It is credibly held that at least three equations are needed for chaotic behavior and it is not surprising that it has been found in the stirred tank when two reactions are taking place. Rössler, Varma and Kahlert (63) have found it for consecutive reactions, one exo- and the other endothermic; Lynch and others (64) have used parallel reactions; the exothermic sequence

$A \rightarrow B \rightarrow C$ has also been considered by Jorgensen (65). This last is a system of three equations for u and v, the concentrations of A and B, and w, the temperature, with seven parameters: α, the Damköhler number for $A \rightarrow B$; β, its dimensionless heat of reaction; γ, its Arrhenius number; κ, a dimensionless heat transfer coefficient; ν, the ratio of activation energies of the two reactions; ρ, the ratio of the heats of reaction; σ, the ratio of the Damköhler numbers. They contain a characteristic non-linearity

$$E(w) = \exp\{\gamma w/(\gamma + w)\}$$

(which reduces to expw if $\gamma \rightarrow \infty$) and are:

$$\dot{u} = 1 - u\{1 + \alpha E(w)\}$$

$$\dot{v} = \alpha u E(w) - v\{1 + \sigma\alpha E^{\nu}(w)\}$$

$$\dot{w} = -(1+\kappa)w + \alpha\beta u E(w) + \alpha\beta\rho\sigma\, v E^{\nu}(w)$$

Jorgensen's first endeavour was to map out the regions of multiplicity, stability and in particular the Hopf bifurcation loci. Then taking all parameters except κ, the cooling rate parameter, to be constant she varied this over an interval ($5 \leq \kappa \leq 8$) in which there was a single unstable steady state and therefore at least one oscillatory solution, the principal oscillation. There were two subintervals, (5.49,5.60) and (6.95,7.29), within which two intricately complex patterns of transitions could be seen, the patterns being topological mirror images of each other. The greater part of each interval was occupied by a cycle of approximately twice the period of the principal oscillatory solution, but then as the values of κ moved outward through the last fractions of the subintervals, cascades of period doublings led to chaos, which subsided into regimes of oscillations of six times the period of the principal. These cascaded into chaos, recovered as period 10 oscillations, relapsed and recovered as period 8's before their final widdershins as fandangos for chaotic and periodic solutions.

"Morris antics" (66) indeed and all within the fifth decimal place! "So what", says the practically-minded engineer: "stultitia est enim illi" (67). But nevertheless it moves and it is not so many years since Amundson's criteria for stability were thought to be idle sophistries. There may indeed be systems for which some of these regions of complexity have a significant size or are situated in an otherwise desirable region of parameter space. As a branch of practical engineering chemical reactors may be tolerably well understood without all the refinements of this intricate behavior. As an object of intellectual study these apparently esoteric features are of great moment; they show that the vein, rich as it has proved to be, is not mined out yet.

The value of comprehensiveness makes demands that open up a question which I believe will be all important in this decade, the question of how to master the daunting "espaces infinis" (68) of a large number of parameters. We need theorems of great power, such as those of Feinberg in kinetics, that will block out whole subspaces and ensure that no exotic type of behavior can hurt or devour there. We need ways of keeping in the important interfaces so that the boundaries of different regions do not have to be determined by trial and error. It is fairly easy to see how to keep on the margin of stability, but how can we tread the edge of chaos? Even this, however, only reduces the dimensionality by one or two and enormous difficulties still remain. Practicality can sometimes come to our aid by confining attention to subspaces which are, in some sense, optimal or, in some sense, safe. But ultimately it will require some new insight to expand our imaginations, revealing the power of a subject apparently so humdrum as chemical reaction engineering to charm "the wide casements opening on the foam of keelless seas in fairy lands folorn" (69).

Literature Cited

1. Baker, H. F., Ed.; "The Collected Mathematical Papers of James Joseph Sylvester" (four volumes); Cambridge, 1912 (biographical notice in Vol. 4).
2. There is some confusion over the reason for Sylvester's untimely departure from the University of Virginia. According to E. T. Bell (Men of Mathematics) he resigned after three months on account of the "refusal of the University authorities to discipline a student who had insulted him." According to Woolf "within three months he was on his way back to England, convinced that he had killed a student in self defense—a conviction that fortunately proved to be false". Of his experiences in 1841 H. F. Baker writes in the Biographical Notice in Volume 4 of the Collected Mathematical Papers; "Such a considerable change deserved a better fate than befell; in Virginia at this time the question of slavery was the subject of bitter contention, and Sylvester had a horror of slavery. The outcome was his almost immediate return; apparently he had intervened vigorously in a quarrel between two of his students."
3. Woolf, H., "An Introduction to J. J. Sylvester"; in "Algebraic Geometry"; Igusa, J-I., Ed.; the Johns Hopkins Centenial Lectures: Supplement to the American Journal of Mathematics.
4. "The Creed of the Old South"; Atlantic Monthly Jan. 1892, 75.
5. "Studies in Honor of Basil L. Gildersleeve"; The Johns Hopkins Press, Baltimore, 1902.
6. Gildersleeve, B. L., "Essays and Studies—Educational and Literary"; Baltimore, 1890.
7. Miller, C. W. E., Ed.; "Selections from the Brief Mention of B. L. Gildersleeve"; Johns Hopkins Press, Baltimore, 1930.

8. Plutarch, De Plac. Phil. i, 5.
9. Pliny, Hist. Nat. ii, 3.
10. Cicero, De Universo, 10.
11. Livingstone, R. W., "A Defense of Classical Education"; Macmillan: London, 1916.
12. Ibid. p. 223.
13. Aris, R., Review of "Insights into Chemical Engineering-Selected Papers of P. V. Danckwerts"; Chem. Engng. Sci. 1982, 37, 1123.
14. Thucydides, Peloponesian War III.82-3.
15. Finley, J. H., "Four Stages of Greek Thought"; p. 73.
16. Gow, A. S. F., "Housman, A. E. A Sketch"; Cambridge University Press: Cambridge, 1936.
17. Times Literary Supplement. 9 May 1968.
18. Housman, A. E., "The Confines of Criticism"; Carter, J., Ed.; Cambridge University Press: Cambridge, 1969.
19. Shakespeare, W., "Twelfth Night"; I, iii, 99-101.
20. Hardy, G. H., "A Mathematician's Apology"; Cambridge University Press: Cambridge, 1940.
21. Whitehead, A. N., "Science and the Modern World"; Macmillan: London, 1925.
22. Ingleton, A. W., Bull. Inst. Math. Applics. 1982, 18, 34.
23. Hoskins, R. F., Bull. Inst. Math. Applics. 1982, 18, 49.
24. Cutland, N. J., Bull. Inst. Math. Applics. 1982, 18, 52.
25. Hardy, G. H., "Ramanujan—Twelve Lectures Suggested by His Life and Work"; Cambridge University Press: Cambridge, 1940.
26. Russell, B. A. W., "The Study of Mathematics" in "Philosophical Essays"; Longmans Green: London, 1910.
27. Wei, J., This observation, despite its interest and importance, seems to have escaped the public record.
28. Aris, R., "Reflections on Some Trends in Chemical Reaction Engineering"; Chap. 4 of "Chemical Engineering in the University Context"; University of Wisconsin Press: Madison, 1982.
29. Damköhler, D., Einflüsse der Strömung, Diffusion und Wärmeiiberganges auf die Leistung von Reacktions öfen, Z. Electrochem. 1936, 42, 846; 1937, 43, 1,8; 1938, 44, 240.
30. Thiele, E. W., Ind. Eng. Chem. 1939, 31, 916.
31. Bischoff, K. B., Ind. Eng. Chem. Fundamentals 1976, 15, 229.
32. Wei, J. and Prater, C. D., "The Structure and Analysis of Complex Reaction Systems". Advances in Catalysis. Academic Press: New York; Vol. 13.
33. Aris, R., "The Mathematical Theory of Diffusion and Reaction in Permeable Catalysts"; Clarendon Press: Oxford; Vol. I, Ch. 5.
34. Amundson, N. R. and Bilous, O., A.I.Ch.E. J. 1955, 1, 513.
35. Wicke, E., "Physical Phenomena in Catalysis and in Gas-Solid Surface Reactions"; Proc. 1st Int. Conf. on Chem. Reac. Engng., Washington, 1970; American Chemical Society: Washington, D.C., 1972.

36. Rony, P. R., J. Am. Chem. Soc. 1972, 94, 8247.
37. Petersen, E. E. and Hegedus, L. L., Ind. Eng. Chem. Fundamentals 1972, 11, 579.
38. Carberry, J. J., Ind. Eng. Chem. 1964, 56, 39.
39. Oh, Se H., Hegedus, L. L., Baron, K. and Cavendish, J. C., "Carbon Monoxide Oxidation in An Integral Reactor. Transient Response to Concentration Pulses in the Regime of Isothermal Multiplicities"; Proc. ISCRE5 ACS Symposium Series 65, American Chemical Society: Washington, D.C., 1978, p. 461.
40. Wei, J. and Degnan, T. F., "Monolithic Reactor-Heat Exchanger"; Proc. ISCRE5 ACS Symposium Series 65, American Chemical Society: Washington, D.C., 1978, p. 83.
41. Gavalas, G. R., "Nonlinear Differential Equations of Chemially Reacting Systems"; Springer-Verlag: Heidelberg, 1968.
42. Amundson, N. R. and Luss, D., A.I.Ch.E. J. 1967, 13, 759.
43. Luss, D. and Balakotaiah, V. Chem. Engng. Sci. (to appear).
44. Danckwerts, P. V., "Insights into Chemical Engineering (Selected Papers of P. V. Danckwerts)"; Pergamon Press: Oxford, 1982.
45. I have touched on some aspects of this in the Wilhelm Lecture for 1981 and in 24.
46. Feinberg, M., Chemical Oscillations, Multiple Equilibria and Reaction Network Structure in "Dynamics and Modelling of Reactive Systems"; Stewart, W. E., Ray, W. H. and Conley, C. C., Eds.
47. Golubitsky, M. and Keyfitz, B. L., SIAM Journal Math. Anal. 1980, 11, 216.
48. Balakotaiah, V. and Luss, D., A.I.Ch.E. J. (to appear).
49. Humphrey, A. E. and Aris, R., Biotech. and Bioeng. 1977, 19, 1375.
50. Aris, R., "Mathematical Modelling Techniques"; Pitman: London, 1978.
51. Aris, R., Chem. Eng. Educ. 1976, 10, 114.
52. Aris, R., Chem. Engng. Sci. 1969, 24, 149.
53. The Watkins number is a dimensionless measure of appropriateness in the closed interval $0 \leq Wa \leq 1$. It was introduced by Re in "a conversation on some aspects of mathematical modelling"; Appl. Math. Modelling 1977, 1, 386.
54. Amundson, N. R. and Aris, R., Chem. Engng. Sci. 1958, 1, 122.
55. Denbigh, K. G., Trans. Faraday Soc. 1944, 40, 352.
56. Uppal, A., Ray, W. H. and Poore, A. B. Chem. Engng. Sci. 1974, 29, 967.
57. Vaganov, D. A., Samoilenko, N. G. and Abramov, V. G., Chem. Engng. Sci. 1978, 33, 1133.
58. Sheintuch, M. and Schmitz, R. A., Cat. Rev. Sci. and Eng. 1977, 15, 107.
59. Sheintuch, M., Kinetic Instabilities in Catalytic Oxidation Reactions (a review to appear).
60. Lorenz, E. N., J. Atmos. Sci. 1963, 20, 130.

61. It would be impossible to give an adequate bibliography of the work on chaos and strange attractors, but the proceedings of a Los Alamos Conference held in May 1982 under the title of "Order in Chaos" will, when they appear, give a good picture of the variety of activity. An introductory sketch of one aspect is given in 53.
62. Pismen, L. M., Chem. Engng. Sci. 1980, 35, 1950.
63. Rössler, O., Varma, A. and Kahlert, C. in "Modelling Chemical Reaction Systems"; Ebert, K. and Jaeger, W., Eds., Springer Verlag: Heidelberg, 1981.
64. Lynch, D. T., Rogers, T. D. and Wanke, S. E., Math. Modelling 1982, 0, 000.
65. Jorgensen, D. V. and Aris, R. will be in the Jan. 1983 issue of Chem. Engng. Sci.
66. Bridges, R. W., "The Testament of Beauty. III"; Oxford, 1929, 1. 938.
67. Hieronymus, E., Vulgate version of I. Cor. 2, 14.
68. Pascal, B., Pensées. III, 206.
69. Keats, J., Ode to a Nightingale, u. 69-70. I quote Keats' original version so that the reader may have the enjoyment of discussing its variations from the well-known final text.

RECEIVED March 10, 1983

2

Development of the Methanol-to-Gasoline Process

J. E. PENICK, W. LEE,[1] and J. MAZIUK

Mobil Research and Development Corporation, New York, NY 10017

Mobil has developed a new process for conversion of methanol to high quality gasoline using a unique class of shape-selective zeolite catalysts. This process provides a novel route to gasoline from either coal or natural gas via intermediate methanol. Two reactor systems have been devised by Mobil for this process--fixed bed and fluid bed. The first commercial application of the fixed-bed process will be in a 14,000 B/D (gasoline) plant based on natural gas which will be located in New Zealand. The fluid-bed process is being scaled up to a 100 B/D (methanol) demonstration plant located in W. Germany. In this paper, the chemistry of the process is reviewed and the engineering involved in selection, scale-up, and design of an appropriate reactor system to produce an acceptable product is discussed.

Mobil has developed a novel process for converting methanol into high quality gasoline.(1) Since methanol can be made from natural gas or coal by well-established commercial technology, the Methanol-to-Gasoline (MTG) process provides a unique approach to the production of gasoline from either raw material as shown in Figure 1. In the MTG process, methanol is quantitatively converted to hydrocarbons and water. The hydrocarbons are primarily gasoline boiling range materials suitable for use as high quality automotive fuel.

[1] Correspondence should be sent to W. Lee, Mobil Research and Development Corporation, Paulsboro, NJ 08066

0097–6156/83/0226–0019$08.50/0

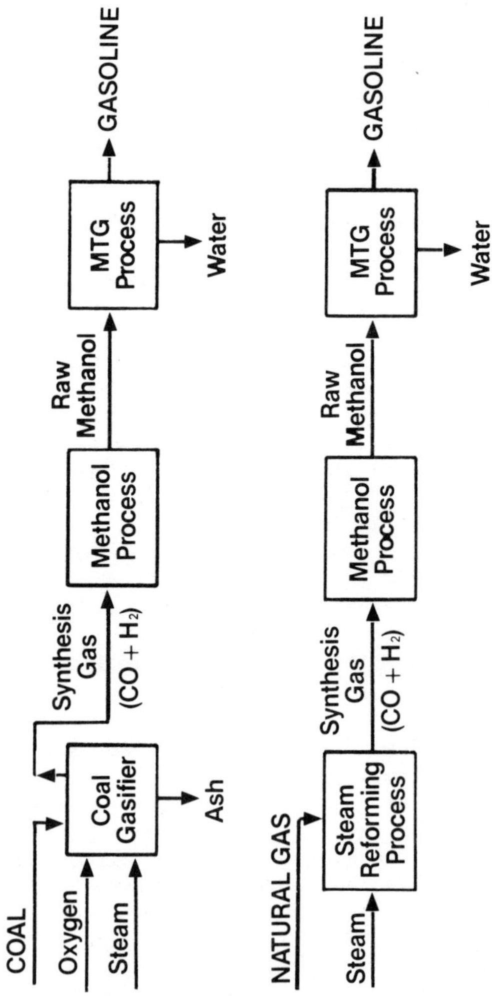

Figure 1. Making gasoline from coal or natural gas via MTG.

The MTG process will soon see its first commercial application in a plant currently being constructed for the New Zealand Synthetic Fuels Corporation Limited. This plant, scheduled for completion in 1985, will produce gasoline from methanol derived from New Zealand natural gas. It will produce some 570,000 tonnes per year (14,000 BPSD) of unleaded gasoline averaging 92 to 94 research octane. This is equivalent to about 1/3 of New Zealand's gasoline consumption. Two 2,200 tonnes-per-day methanol plants will provide the feed for the single-train MTG plant.

In this lecture, the development of the MTG process will be reviewed. First, the unique aspects of MTG -- the catalyst, chemistry, and its special reactor design aspects -- will be discussed. Next, the choices for the conversion system will be presented along with the dual-pronged strategy for development of both the fixed- and fluid-bed processes. Finally, our future development plans for this general area of technology will be highlighted.

Catalyst

The ZSM-5 catalyst, the key element in the MTG process, is a shape-selective synthetic zeolite with a unique channel structure as shown in Figure 2.(2) This structure is distinctively different from the familiar wide-pore zeolite, faujasite, and the narrow-pore zeolites. The lines in Figure 2 represent oxygen atoms in a siliceous framework. It contains a novel configuration of linked tetrahedra consisting of eight five-membered rings. These units join through edges to form chains and sheets, leading to the three dimensional structure representation shown in the figure. It contains two intersecting channels: elliptical 10-membered ring straight channels and sinusoidal, tortuous channels. The pore-opening of these channels is about 6 Å and is just wide enough to produce hydrocarbons boiling primarily in the gasoline range.

Since the mid-1970's, more than 25 commercial plants have been commissioned using this catalyst family. There are currently five licensed processes (other than MTG) based on the ZSM-5 family:

- Distillate dewaxing
- Xylene isomerization
- Toluene disproportionation
- Ethylbenzene synthesis
- Lubricating oil dewaxing

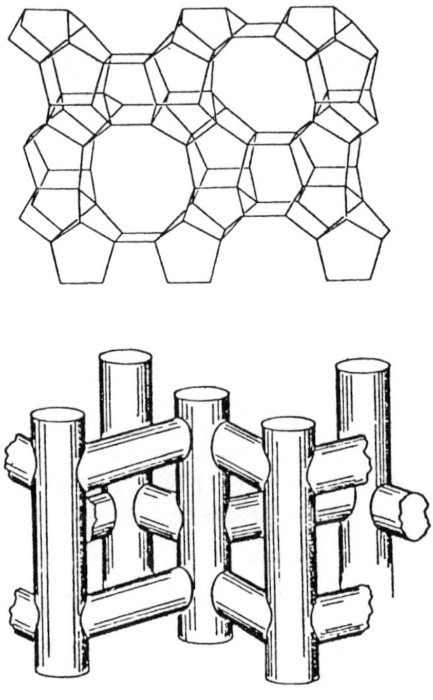

Figure 2. ZSM-5 structure. (Reproduced with permission from Ref. 2. Copyright 1978, Macmillan.)

All of these applications use fixed-bed reactors, and, very importantly, all were scaled up from bench-scale pilot plant data. This successful scale-up experience with the ZSM-5 catalyst was an important consideration in formulating the MTG development strategy.

Chemistry

The conversion of methanol to hydrocarbons over ZSM-5 is well-represented by the following simplified reaction network(3):

$$2CH_3OH \underset{+H_2O}{\overset{-H_2O}{\rightleftharpoons}} CH_3OCH_3 \xrightarrow{-H_2O} \underset{\text{Olefins}}{C_2\text{-}C_5} \longrightarrow \begin{matrix}\text{Paraffins}\\ \text{Aromatics}\\ \text{Cycloparaffins}\end{matrix}$$

The initial dehydration reaction is sufficiently fast to form an equilibrium mixture of methanol, dimethyl ether, and water. These oxygenates dehydrate further to give light olefins. They in turn polymerize and cyclize to form a variety of paraffins, aromatics, and cycloparaffins. The above reaction path is illustrated further by Figure 3 in terms of product selectivity measured in an isothermal laboratory reactor over a wide range of space velocities.(3) The rate limiting step is the conversion of oxygenates to olefins, a reaction step that appears to be autocatalytic. In the absence of olefins, this rate is slow; but it is accelerated as the concentration of olefins increases.

Under MTG conditions, almost no hydrocarbons are found higher than C_{10} due to the shape selectivity of ZSM-5. While higher molecular weight compounds may be formed within the catalyst channels, they do not escape due to the spatial hindrance of the catalyst structure.

Reactor Design

Although the design of a reactor system for the MTG process involves classical chemical engineering principles, the unique catalyst and reaction mechanisms impose important design constraints. These include the highly exothermic nature of the reaction, the need for essentially complete methanol conversion, steam deactivation of the catalyst, the "band-aging" phenomena, and durene formation.

- Exotherm - The heat of reaction is 1740 kJ/kg of methanol which if uncontrolled would give an adiabatic temperature rise of about 600°C. Therefore, a principal consideration in designing a reactor system is reaction heat management.

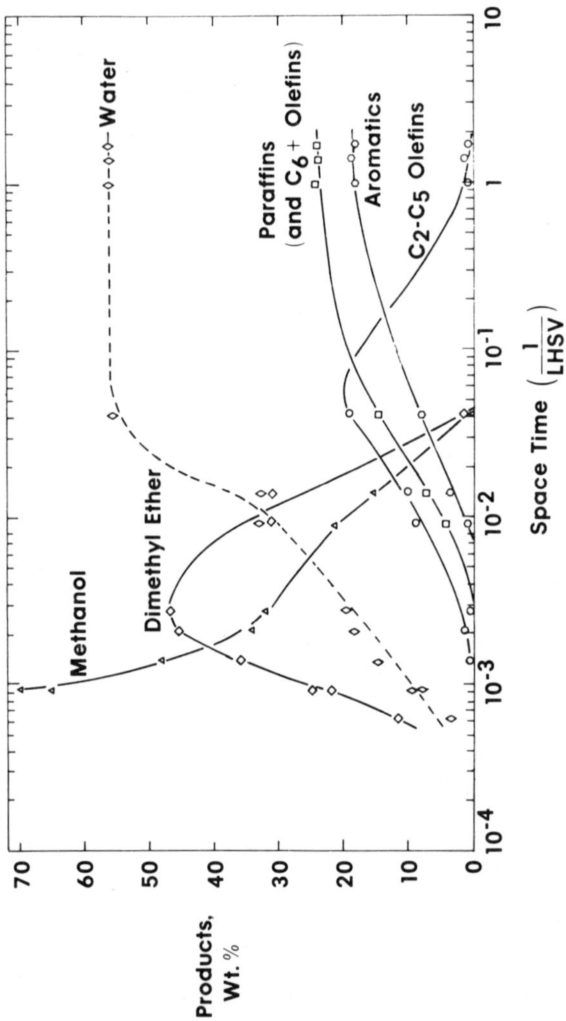

Figure 3. Effect of space velocity at 101 kPa and 371 °C. (Reproduced with permission from Ref. 3. Copyright 1977, Academic Press.)

- Complete Methanol Conversion - The major products of the MTG conversion are hydrocarbons and water. Consequently, any unconverted methanol will dissolve into the water phase and be lost unless a distillation step to process the very dilute water phase is added to the process. Thus, essentially complete conversion of methanol is highly preferred.

- Catalyst Deactivation - Under MTG reaction conditions, ZSM-5 catalyst undergoes two types of aging which contribute to a gradual loss of catalyst activity. A reversible loss results from "coke" formed on the catalyst as a reaction byproduct. This deactivation is a typical catalytic process design problem. What is unusual about the MTG process is that a reaction product, steam, is also responsible for a gradual loss of activity. However, low reactor temperatures and low partial pressure of water will minimize this aging and favor a long catalyst life.

- Band-Aging - Especially with fresh catalysts, the reaction occurs over a relatively small zone in a fixed bed. This reaction front marches down the catalyst bed as the coke deposits first deactivate the front part of the bed (Figure 4). Use of a sufficient catalyst volume permits a fixed-bed design in which on-stream periods are long enough to avoid overly frequent regeneration cycles.

- Durene Formation - An aromatic compound, durene (1, 2, 4, 5-tetramethylbenzene) is produced in the MTG process. Durene has excellent research octane blending quality (110 RON clear) and boils within the gasoline distillation range (197°C); however, its freezing point is relatively high at 79°C. Based on our current designs, the gasoline directly from the ZSM-5 reactors could contain 4 to 7 wt% durene. These concentrations of durene could lead to problems with solids build-up in carburetors during cold starts. While durene is present in conventional gasoline, the typical concentrations are low enough not to cause problems.

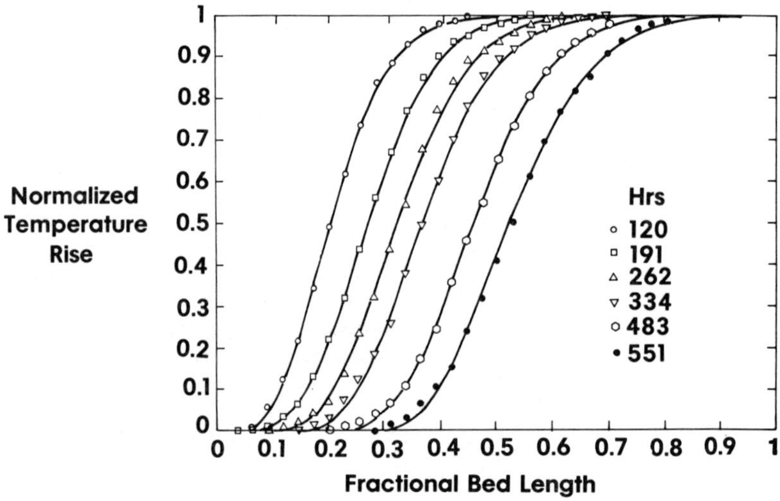

Figure 4. Movement of normalized temperature profile through bed.

In the MTG process, durene is mostly formed by alkylation of lower molecular weight aromatics with methanol and/or ether. Low methanol partial pressures and high reaction temperatures tend to reduce the durene level, presumably by reducing the concentration overlap of methanol/ether with the aromatics formed. This overlap tends to increase with larger catalyst particles. As a consequence the early fixed-bed development work was conducted using 1 mm diameter extrudates.

All of the above factors must be considered in the various reactor systems for the MTG process: fixed-bed reactors, fluid-bed reactors and tubular reactors. Figure 5a shows three varieties of fixed-bed reactors and Figure 5b illustrates heat-exchanger and fluid-bed reactor variants.

Fixed-Bed Systems

The major advantage of these fixed-bed systems is that the fixed-bed technology can be simple, and will require minimum scale-up studies. As indicated earlier, this type of design has been successfully scaled-up to commercial size directly from bench-scale pilot plant data many times.

One common method for managing the adiabatic temperature rise in fixed-bed reactors is to dilute the reactant stream with a gas that would provide the mass to absorb the heat of reaction. Light hydrocarbon reaction products (methane, ethane, propane) can be recycled for this purpose. However, the costs associated with the recycle operation could be substantial if the reaction is strongly exothermic. Special attention must be given to reduce the recycle ratio.

Use of multiple beds/reactors in series with intercooling or quenching is a method which can be used to reduce the amount of recycle and its associated costs. Multiple catalyst beds reduce costs by using the recycle material several times before it is separated from the reaction products.

In Figure 6a, a multibed reaction system is shown for which only partial conversion is taken across each bed; the effluent is cooled before entering each succeeding bed such that the effluent from each bed does not exceed a specified maximum temperature. The band-aging phenomena previously discussed would complicate the development of such an MTG process since the catalyst appears to age from the front to the back. Thus, it might be difficult, especially for the first bed, to maintain partial conversion at a given level for a reasonable amount of time.

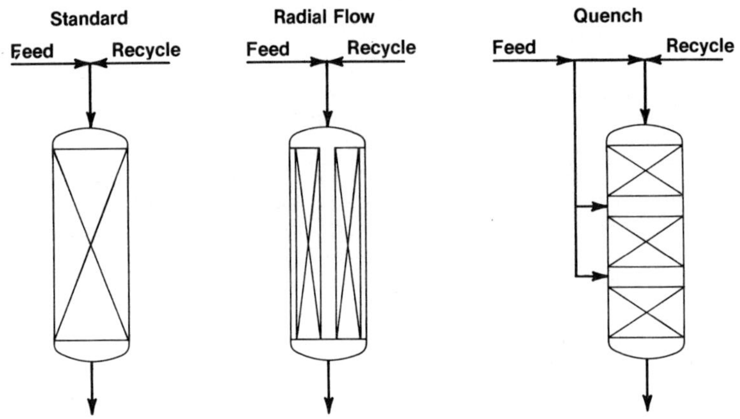

Figure 5a. Adiabatic reactors.

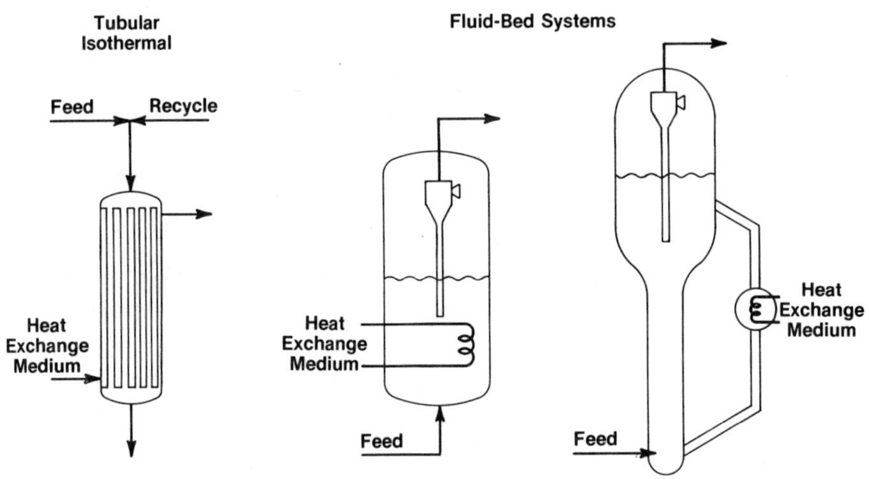

Figure 5b. Nonadiabatic reactors.

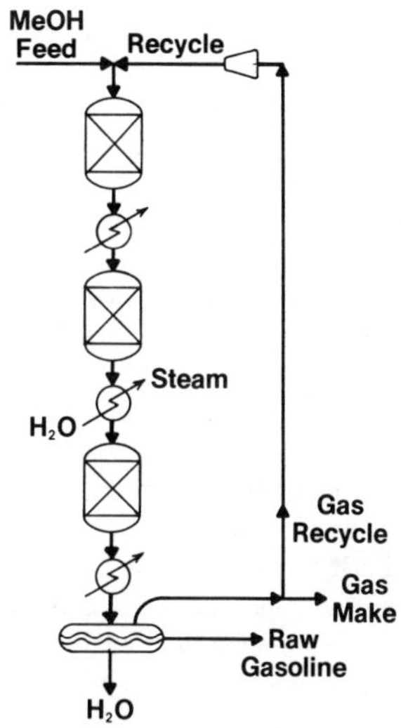

Figure 6a. Multibed designs. Interstage heat removal.

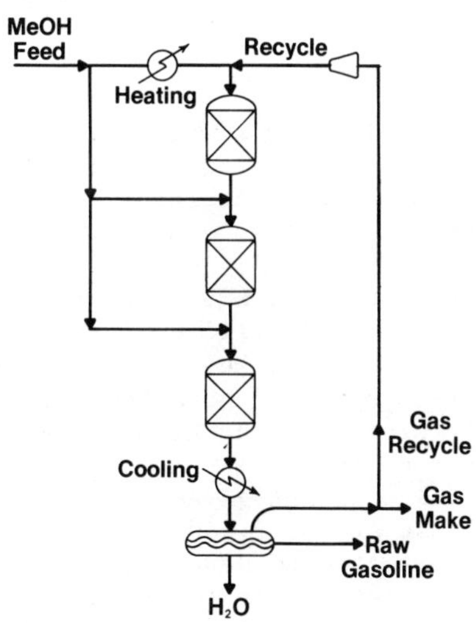

Figure 6b. Multibed design. Interstage feed quenching.

In Figure 6b, a multibed reaction system is shown in which only part of the reactant is fed to each bed. The reactant fed to each bed is essentially completely converted and the reaction products from that conversion become part of the temperature-moderating material for each succeeding bed. In such a reaction system the methanol feed to each succeeding bed would be mixed with aromatics made in the preceding beds - - an arrangement that tends to increase durene formation.

Still another multi-reactor approach is to divide the MTG reaction into two steps as shown in Figure 7. In the first step, methanol is partially dehydrated to form an equilibrium mixture of methanol, dimethyl ether and water over a dehydration catalyst. About 15% of the reaction heat is released in this first step. In the second step, this equilibrium mixture is converted to hydrocarbons and water over ZSM-5 catalyst with the concomitant release of about 85% of the reaction heat. Though this two step approach does not have any of the inherent complications of the previously mentioned multibed reaction systems, it leaves one with a substantial amount of the reaction heat (85%) still to be taken over one catalyst bed. This requires a fairly high recycle stream to moderate the temperature rise over the second reactor. Such a high recycle design would require careful engineering in order to transfer heat efficiently from the reactor effluent to the recycle gas and reactor feed. However, this two stage reactor system is the simplest of the fixed-bed systems to develop.

Fluid-Bed Reactor Systems

A reactor system which has several attractive features for heat removal is the fluid bed. With a fluid bed, the reaction heat can be removed by using heat exchange coils in the bed, or catalyst can be circulated to an external cooler for heat recovery. Heat transfer in a fluid bed is so rapid that the danger of local overheating is minimal. Furthermore, the heat of reaction can be recovered efficiently as high pressure steam. As expected, the fluid-bed yield pattern is different from that of the fixed bed; C_5+ gasoline yield is lower, but the light olefins yields are much higher, partly because light product gas is not recycled. Alkylation of these olefins with product isobutane and blending gives an increased gasoline yield compared to that of the fixed bed.

Two factors complicate the development of a fluid-bed MTG process. One is the need to ensure the complete conversion of methanol; the other is to develop a fluid catalyst sufficiently rugged to withstand the abrasive forces inherent in fluid-bed operation.

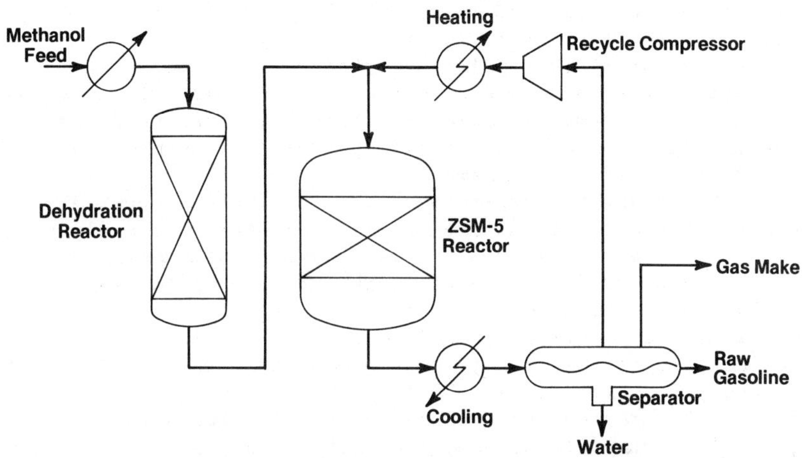

Figure 7. Two-stage fixed-bed MTG.

Since in a large diameter fluid bed, bubbles may bypass the catalyst, elaborate hydrodynamic studies are required to scale-up the reactor internals to accomplish the complete conversion of methanol. As mentioned earlier, essentially complete conversion of methanol is a process requirement; otherwise, a costly distillation step must be added to recover unconverted methanol.

Thus, the fluid-bed MTG process was assessed as being a desirable but a long-term development. The scale-up would have to proceed through several stages before it would be ready for commercialization.

Tubular, Heat-Exchange Reactor Systems

Another reactor system which has several attractive features for heat removal is the tubular, heat-exchange reactor. Good temperature control can be achieved in the tubular reactor if the coolant approximates an isothermal heat sink. Light gas recycle can be reduced significantly compared to fixed-bed systems. Tubular reactors have been used for Fischer-Tropsch reactions and for synthesis of methanol and phthalic anhydride, for example.

The MTG reaction temperature of approximately 400°C essentially precludes the use of boiling water as a coolant due to the high steam pressures involved. While oil cooling might conceivably be used, the temperatures are above the range usually considered practical for such a design and large flow rates would be required to achieve uniform cooling. Heat transfer salt and boiling mercury are other possibilities; each presents particular design challenges. Of these, heat transfer salt probably represents the best compromise for an MTG application primarily because the salt could also be used to control the periodic catalyst regeneration operation which entails temperatures of about 500°C. From our viewpoint the heat exchange reactor has the following disadvantages: (1) molten-salt cooling presents significant operational problems, (2) tubular reactors generally contain less catalyst per unit of vessel volume than other reactor types, (3) the maximum practical capacity for a single reactor is less than for other types and (4) very careful mechanical design will be required for a tubular reactor which can reliably withstand both the conversion and catalyst regeneration cycles.

Strategy for Commercialization

Since there are many possible reaction systems for an MTG process, it was necessary to focus the development efforts. In the mid-1970's, Mobil decided to pursue a double-pronged approach in order to ensure both short- and long-term objectives. The plan was:

NEAR-TERM --Proceed with commercialization of a simple, direct, and uncomplicated fixed-bed technology. Such a development would parallel extensive experience in successful commercialization of many fixed-bed processes using similar catalysts and operating conditions. The five licensed processes using ZSM-5 catalyst fall in that category. The simplest fixed-bed MTG system was the one which employed dehydration and ZSM-5 reactors. This system was studied extensively in bench-scale units. These studies in a 3 cm diameter by 30-50 cm length reactors were considered to be sufficient for scale-up.

However, in 1978, the New Zealand government expressed strong interest in applying the MTG technology for conversion of indigenous natural gas to high-quality gasoline. In support of the New Zealand effort, it was decided to demonstrate the fixed-bed process in a 4 B/D pilot plant. Such a demonstration test could be fit into the overall schedule for the New Zealand venture and would provide large quantities of product gasoline for road testing in cars typically sold in New Zealand.

LONG-TERM --Proceed with an orderly development of the fluid-bed process, which may offer an attractive alternative to the fixed-bed process for the future with its possibly higher gasoline yield, more efficient heat recovery, and economy of scale. Currently, a 100 B/D fluid-bed pilot plant in Germany is nearing start up. This study is a joint project with URBK, Uhde, and Mobil supported in part by the U.S. DOE and the German government. Successful completion of the 100 B/D program will be a prerequisite for proceeding with commercialization of the fluid-bed process.

Fixed-Bed Development and Commercialization

Fixed-Bed Demonstration Plant. The major objective of the demonstration test was to verify the bench unit results in a larger plant operating at conditions similar to a commercial-size reactor. The linear velocity of the reactant is the only variable which is significantly different between a bench-scale unit and a commercial-size reactor. The other operating conditions are normally the same for both reactors. By making the length of the catalyst bed in the demonstration plant identical to that of the proposed commercial-size reactor, the gas velocities also become identical, and thus all the process variables become identical.

For a simple gas recycle system, there is a strong incentive to have the bed length be as short as possible to hold down the pressure drop and thus compression costs. But as catalyst beds get short, one has to be concerned with maldistributions in the catalyst packing and consequently in the flow through the bed. If there were regions of high flow in an MTG catalyst bed, unconverted methanol could breakthrough early, resulting in either short cycles or in methanol loss to the water phase. Mobil came to the conclusion that with a reactor bed length of 2.5 meters, it could effectively load the catalyst and ensure even distribution of the gas in reactors having large diameters.

The catalyst bed length for the demonstration test was set at the likely length for a commercial plant. For the demonstration unit, the internal diameter was selected to be about 10 cm, which is significantly larger than catalyst particle size to minimize wall effects. In addition, heat transfer along the reactor wall would also be negligible.

The results of the demonstration plant are compared with those of the bench unit in Table I. The two units were operated at identical conditions except for gas velocity. The product yields and selectivities over the first cycle were remarkably similar. The adiabatic temperature rise and the band aging behavior were also the same. However, the demonstration unit performed considerably better with respect to cycle length. Typical cycles were about 50% longer. Similar behavior of longer cycle lengths have been observed in commercial operation with other ZSM-5 catalyzed processes.

In the product quality tests, particular attention was given to the effect of durene on automobile performance and gasoline handling since its content in the MTG gasoline is the only significant difference from conventional gasolines. A test fleet designed to simulate the New Zealand automobile population was used for the driveability tests. Figure 8 illustrates the results of one test series conducted at -18°C ambient temperature.(4) The data show the weight percent durene that could be tolerated before encountering any abnormal driveability characteristics. All cars performed normally on gasoline containing 3 wt% or less durene. With 4 wt% durene, the most critical car encountered a slightly rougher-than-normal engine idle condition. Of the 12 cars tested at -18°C, eleven had no decrease in driveability at 4 wt% durene, 7 were good at 5 wt%, and 3 were good at 6 wt% durene. Much higher durene tolerances were observed at higher temperatures.

To ensure that there would be no driveability or handling problems with MTG gasoline, a target of 2 wt% durene in gasoline was set as the company specification, though the MTG

Table I. Scale-Up Results of Fixed-Bed MTG

	Bench Unit	4 B/D Unit
Conditions		
Methanol/Water Chg. (W/W)	83/17	83/17
Conversion Reactor Inlet Temperature (°C)	358	360
Conversion Reactor Outlet Temperature (°C)	404	407
Methanol Space Velocity (WHSV)	1.6	1.6
Separator Temperature (°C)	49	52
Recycle Ratio (mol/mol chg)	9	9.2
Pressure (kPa)	2,163	2,156
	Average First Cycle Yield (Wt%)	
Hydrocarbon Products		
Methane	1.33	1.25
Ethane	0.82	0.86
Ethylene	0.02	0.03
Propane	8.54	8.60
Propylene	0.15	0.15
Isobutane	8.45	8.39
n-Butane	4.06	4.20
Butenes	0.71	0.74
C_5+ Hydrocarbons	75.92	75.78
	100.00	100.00
Gasoline (Wt%)	80.2	80.2
Octane, R+0	95	95

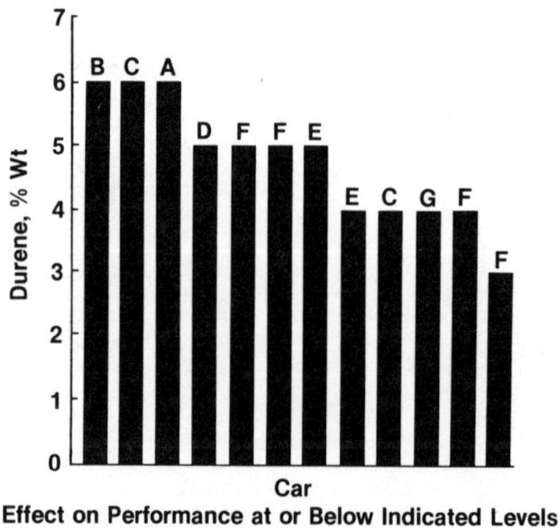

Figure 8. Durene tolerance of automobiles at -18 °C.

process was producing about 5 wt% at that time. To meet this specification with conventional processing would have required undercutting the gasoline product.

Consequently, an exploratory program was undertaken to see if this specification could be met by addition of a simple gasoline treating step. In a short time, a heavy gasoline treating (HGT) technology evolved. With the HGT, durene levels even higher than 5 wt% could be treated down to the 2% level.

The HGT efficiently fits into the processing scheme. The hydrocarbon liquid from the separator (shown in Figure 9) is normally stabilized by fractionating at around 140 psig; this pressure produces a relatively hot gasoline bottoms stream. Sending this high heat content stream directly to a low-pressure gasoline splitter allows us to concentrate the durene in a small bottoms stream. Only small amounts of energy need to be expended since the feed is so hot and the separation can be made with little reflux.

The small amount of hydrogen required for the HGT can be readily obtained from a portion of the methanol synthesis purge gas. This purge stream is normally used as fuel; the fuel value of the hydrogen can be replaced by the small amount of gas produced from the HGT step.

The bottom portion of the gasoline (approximately 15% in the NZ design) is processed in a catalytic reactor designed to convert the durene by both isomerization and dealkylation. The treated heavy gasoline is then stabilized and reblended with the light gasoline. Neither the finished gasoline volume nor its octane numbers are affected. The minor weight loss from the treating step is offset by a decrease in the specific gravity of the product. Incorporation of the HGT process has greatly eased the durene constraint originally imposed on the MTG reactor design. In fact, as a consequence of incorporating HGT, we are able to switch to a larger catalyst particle size, which makes more durene than the smaller, original catalyst. Such a change brings about less reactor pressure drop, and thus overall, a more efficient plant.

Commercial Fixed-Bed Plant Design. The commercial fixed-bed MTG plant is very similar in design concept to the 4 B/D demonstration plant. A typical design of the reaction section of a commercial plant is shown in Figure 10. The feedstock may be distilled or crude methanol.

The feed is vaporized by reactor effluent heat-exchange and enters into the dehydration reactor. The dehydration reactor effluent is mixed with preheated recycle gas and enters the conversion reactors. Although Figure 10 shows four parallel or "swing" conversion reactors, a lesser or greater number of reactors may be used depending upon the feed rate and regeneration frequency desired.

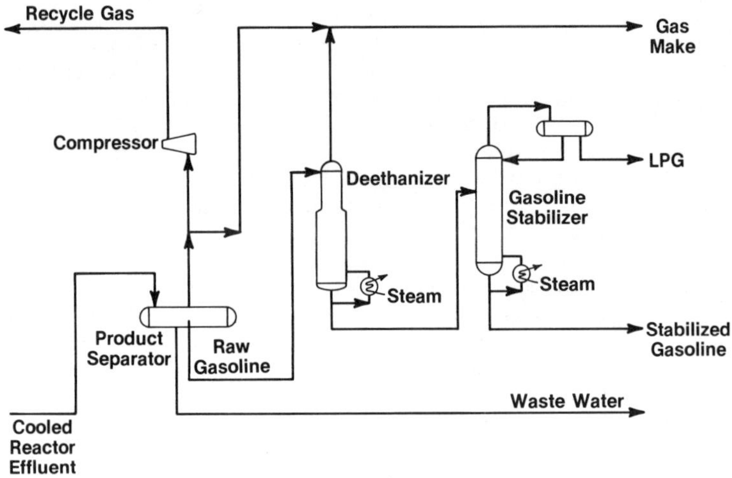

Figure 9. Fixed-bed MTG separations section.

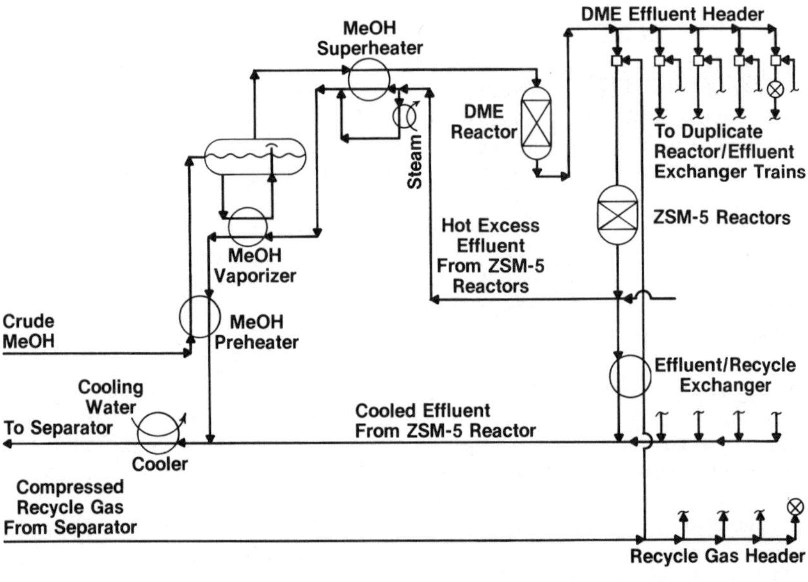

Figure 10. Flowsheet of the reaction section of the fixed-bed MTG process.

We have found that to efficiently recover the high heat content in the reactor effluent it is desirable to split the stream into two parts. The majority of the effluent from a particular reactor is used to heat the recycle gas to that reactor. The excess hot effluents from all the reactors are combined and used for generating steam and heating the methanol feed to the dehydration reactor. After these services, the reactor effluents are combined and cooled by cooling water before entering the product separator.

Considerable engineering judgment and effort are needed to ensure that the heat recovery is efficient, yet has low pressure drop. Since a large portion of the heat contained in the reactor effluent has to be transferred back to the cold recycle gas, this exchanger arrangement received our special attention.

We have also found it advisable to vaporize the methanol using a vapor-liquid drum, analogous to a steam drum. Such an arrangement increases the loop pressure drop because the excess hot effluent flows through separate exchangers in series for preheating, vaporizing, and superheating the methanol. However, this arrangement has two advantages:

- A blowdown of liquid methanol can be taken if the crude methanol contains any solids.
- As the control valves to the various MTG reactors are adjusted, pressure waves may be transmitted to the methanol exchangers. If one exchanger is used for preheating, vaporizing and superheating the methanol feed, the pressure waves may cause the area used for vaporization to fluctuate thereby leading to potential instability in the vaporization rate.

Gas, liquid hydrocarbon, and water phases are separated in the product separator. Most of the gas is recycled. The net gas make and liquid hydrocarbon streams are sent to conventional petroleum fractionators for separation and stabilization as shown in Figure 9.

When methanol breaks through a catalyst bed, that reactor is taken off stream and the catalyst is regenerated by burning off the deposited coke. Air is used as the source of oxygen to burn off the coke and nitrogen dilution is used to control the peak burning temperature. Most of the reactor effluent gases during regeneration are recycled back to the reactor after water is removed.

The reaction engineering in the MTG plant centered on how deep the catalyst bed had to be to avoid premature methanol break-through. The decision that the minimum bed depth had to be 2.5 meters eliminated radial reactors from consideration.

Fluid-Bed Development

From the small bench-scale tests, the fluid-bed MTG development has proceeded to two phases of scale-up. The initial scale-up was vertical to a 4 B/D capacity, and the current phase is a horizontal scale-up to 100 B/D capacity.

Vertical Scale-Up of Fluid Bed. The fluid-bed process was first scaled up vertically from the 45 cm tall bench-scale reactor to the 760 cm tall 4 B/D reactor shown schematically in Figure 11. The 4 B/D reactor internal diameter was 10 cm.

The operation of this 4 B/D fluid pilot plant was very successful. Essentially complete conversion of methanol was achieved at the design conditions for superficial gas velocities up to 0.5 m/sec. Excellent stability of the operation was demonstrated over a 75-day run. Higher gasoline yield and octane number than for fixed-bed processing were verified (Table II). A typical temperature profile in the reactor is shown in Figure 12. Despite the highly exothermic nature of the reaction and the unusually high length-to-diameter ratio (greater than 75) of the reactor, a very uniform temperature profile was established. There was a complete absence of any troublesome "hot spots". In addition, the transient temperature profiles during heat-up and cool-down were also uniform and stable.

With these encouraging results, it was decided to scale-up the fluid-bed process further. A 100 B/D fluid-bed program has been formulated to demonstrate horizontal scale-up.

Horizontal Scale-Up of Fluid Bed. The 100 B/D pilot plant will provide additional information for design of a commercial-scale, fluid-bed process. Key data yet to be obtained are:

- reactor internal baffle design to maintain complete conversion of methanol in a large diameter fluidized vessel.
- catalyst attrition and ultimate life.

A schematic diagram of the reactor system of the 100 B/D plant is shown in Figure 13. There are three major vessels: reactor, regenerator, and external cooler. The reactor consists of a dense fluid-bed section (60 cm ID x 13.2 m height) located above a dilute phase riser. Two modes will be studied to remove reaction heat:

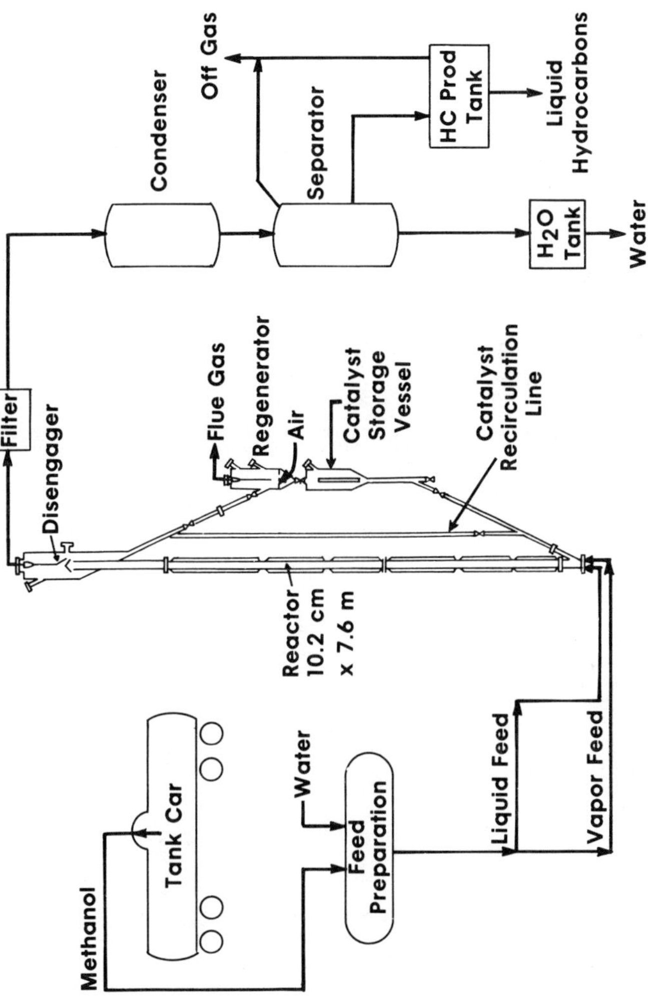

Figure 11. 4 B/D fluid-bed pilot plant.

Table II. Process Conditions and Product Yields from MTG Processes

	Fixed Bed	Fluid Bed
Conditions		
Methanol/Water Chg. (W/W)	83/17	83/17
Dehydration Reactor Inlet Temperature (°C)	316	-
Dehydration Reactor Outlet Temperature (°C)	404	-
Conversion Reactor Inlet Temperature (°C)	360	413
Conversion Reactor Outlet Temperature (°C)	415	413
Pressure (kPa)	2,170	275
Recycle Ratio (mol/mol chg.)	9.0	-
Space Velocity (WHSV)	2.0	1.0
Yields (Wt. % of Methanol Charged)		
Methanol + Ether	0.0	0.2
Hydrocarbons	43.4	43.5
Water	56.0	56.0
CO, CO_2	0.4	0.1
Coke, Other	0.2	0.2
	100.0	100.0
Hydrocarbon Product (Wt. %)		
Light Gas	1.4	5.6
Propane	5.5	5.9
Propylene	0.2	5.0
Isobutane	8.6	14.5
n-Butane	3.3	1.7
Butenes	1.1	7.3
C_5+ Gasoline	79.9	60.0
	100.0	100.0
Gasoline (including Alkylate) [RVP-62 kPa (9psi)]	85.0	88.0
LPG	13.6	6.4
Fuel Gas	1.4	5.6
	100.0	100.0
Gasoline Octane (R+O)	93	97

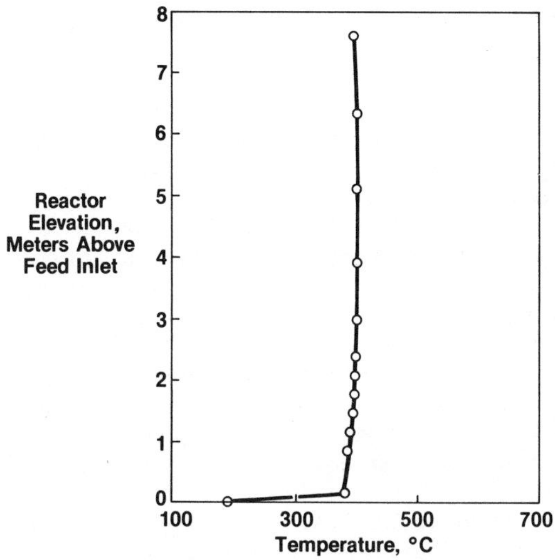

Figure 12. Fluid-bed reactor temperature profile.

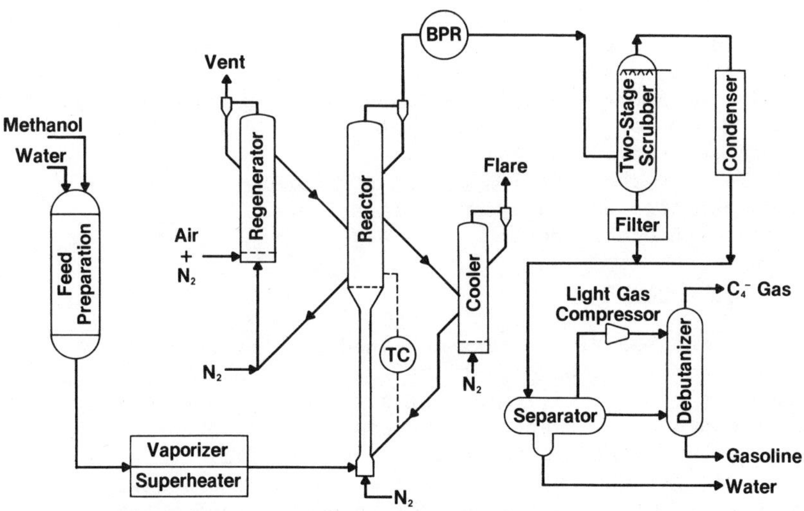

Figure 13. 100 B/D fluid-bed MTG demonstration plant.

- circulation of catalyst through an external catalyst cooler, and

- internal heat exchange pipes immersed in the dense bed reactor.

To maintain a constant catalyst activity in the reactor, a small fraction of "coked" catalyst will be continuously regenerated and returned to the reactor.

The mechanical design basis of the 100 B/D unit has been verified on a full-scale Cold Flow Model (CFM). This non-reacting model proved very useful for optimizing baffle design and catalyst circulation strategies. Several different baffle designs -- horizontal and vertical arrangements (Figure 14) -- have been tested in the CFM using several experimental techniques: gas tracer, capacitance probes, bed expansion analysis, and visual observation. Integrated residence time behavior of the bed and quantitative, local gas/solids distribution and bubble size distribution were measured.

Experimental results indicate that horizontal baffles are effective in breaking bubbles. In the turbulent fluidization regime, horizontal baffling provides also the effect of staging the fluid bed and limiting the formation of gross circulation patterns within the bed. Thus, horizontal baffles were selected for installation in the pilot plant. The baffle design was optimized to ensure sufficient catalyst flux through the baffled section. An overly restrictive design can cause a marked density gradient in the bed, with most of the catalyst accumulating above the top baffle section. The final baffle design chosen for the 100 B/D pilot plant did not exhibit this catalyst segregation phenomenon.

We believe that the 100 B/D plant represents the final development step before a commercial-size reactor can be designed with confidence.

Future Development

The fixed-bed MTG process designed for New Zealand represents a simple but highly reliable design concept. The development of the HGT, which relaxes the durene constraint, opens the process variable window considerably. For example, if the recycle ratio is reduced from 9:1 to 7:1, the size of the recycle circuit is reduced and it becomes more efficient, leading to reductions of about 30% in heat transfer area and compressor horsepower. However, the octane of the gasoline product is reduced by about one number. Several more advanced fixed-bed process arrangements which reduce processing costs without reducing octane are being considered and subjected to preliminary screening tests.

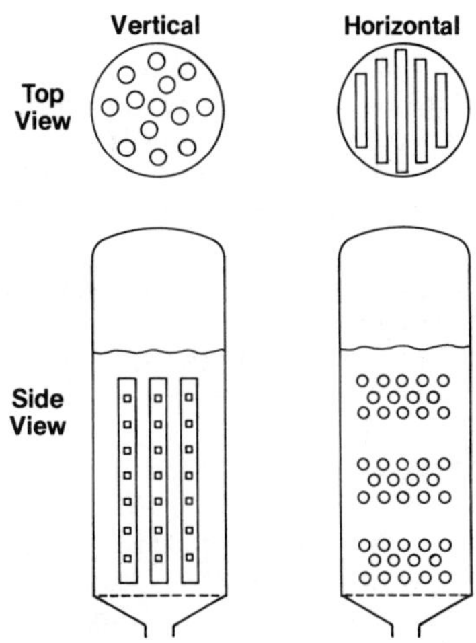

Figure 14. Examples of fluid-bed reactor internal baffles.

The fluid-bed process being developed in the 100 B/D pilot plant has good potential for becoming an important part of our future technology. It has inherent potential advantages over the fixed-bed process as mentioned earlier. It is basically flexible for easy adaptation to petrochemicals production and better control of the autocatalytic reaction.

We are looking at catalyst modifications both for improved yield/octane characteristics and for improved steam stability. It has been our experience that new, improved catalysts are an integral part of reaction engineering. To keep processes vital and competitive, it is necessary to continually look for improved catalysts.

In addition to the production of gasoline, the prospects look promising for manufacturing diesel fuels and chemicals using related technology. Referring back to Figure 3, which shows the reaction steps involved in methanol conversion to hydrocarbons, we observe that light olefins form first as intermediate products. As the reaction proceeds, the light olefins react further to form isoparaffins and aromatics. In the gasoline production mode, we are operating at the far right where the gasoline components dominate.

By modification of catalyst and/or process conditions, one can operate where the selectivity to light olefins is high. We have learned to increase the ethylene yield to about 30 wt% of hydrocarbon, for instance. This process is perhaps the most promising new route for the production of ethylene from synthesis gas.

In our exploratory studies, we have now found good leads for yielding as much as 75-80% C_2-C_4 olefin product, which in turn can be converted to good quality distillate.(5,6) We can now consider the co-production of gasoline and distillate from methanol as shown in Figure 15. Engineering and development studies are in the initial phases of evaluating such concepts.

The zeolite-catalyzed methanol conversion technology, whether the desired product is gasoline, diesel, jet fuel or ethylene for petrochemicals, will provide new opportunities for synfuels in the coming decades.

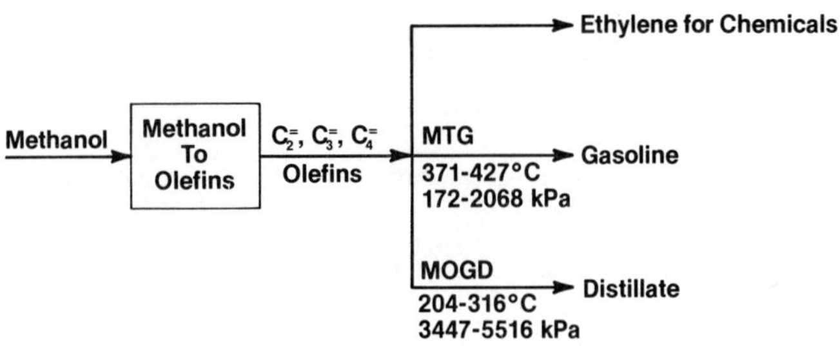

Figure 15. Production of ethylene, gasoline, and distillate from methanol.

Literature Cited

1. Meisel, S. L., McCullough, J. P., Lechthaler, C. H., and Weisz, P. B., CHEMTECH, 1976, 6, 86-9.
2. Kokotailo, G. T., Lawton, S. L., Olson, D. H., and Meier, W. M., Nature, 1978, 272, 437-8.
3. Chang, C. D., and Silvestri, A. J., J. Catal., 1977, 47, 249-59.
4. Fitch, F. B., and Lee, W., "Methanol-to-Gasoline, An Alternative Route to High Quality Gasoline," presented at the International Pacific Conference on Automotive Engineering, Honolulu, Hawaii, Nov., 1981.
5. Garwood, W. E., "Conversion of C_2-C_{10} Olefins to Higher Olefins Over Synthetic Zeolite ZSM-5," presented at Am. Chem. Soc. Mtg., Las Vegas, Nevada, March, 1982.
6. Meisel, S. L. and Weisz, P. B., "Hydrocarbon Conversion and Synthesis Over ZSM-5 Catalysts", presented at Advances in Catalytic Chemistry II Symposium, Salt Lake City, Utah, May, 1982.

RECEIVED March 10, 1983

3

Fluidized-Bed Reactors

P. N. ROWE

University College London, Department of Chemical and Biochemical Engineering, Torrington Place, London WC1E 7JE, England

This Plenary Lecture surveys the place of fluidised beds within the field of chemical reaction engineering and describes some developments and changes of interest over the last decade. It concludes by suggesting the direction basic research should take in the future.

The previous six International Symposia on CRE have included 56 plenary lectures three of which were devoted exclusively to fluidised bed reactors and a further two were substantially concerned with them (1-5). This is a level of interest equalled only by the general subject of optimisation, stability and control and exceeded only be plenary lectures concerned with kinetics (just exceeded; there were six). Here we are with yet another taste of this fascinating, intriguing, challenging, interest-full, promising and versatile form of chemical reactor. There is a lot of juice yet left in such a succulent subject although it becomes increasingly difficult to treat with originality a topic that has been mouthed by so many.

Two years ago we were treated to a nostalgic account of the coming of age of chemical reaction engineering by Professor Levenspiel who dated its birth as 1957 and its possible conception as 1947 (6). Uncharacteristically Octave seems to have overlooked Denbigh's classical paper (7) submitted in 1943 in which Denbigh described and defined the continuous flow tubular reactor and the continuous flow stirred tank reactor for each of which he showed how to calculate both yield and size of reactor needed for given orders of reaction and required production rates. He also showed how selectivity would depend on the order of competing reactions. This and Danckwerts' classical paper on residence time distributions (8) submitted in 1952, laid most of the foundations on which chemical reaction engineering has since built.

It is an interesting experience to re-read these papers, especially Denbigh's. It is quite a long paper (22 pages, he

0097-6156/83/0226-0049$06.00/0

would never get away with that length today) and gives an impression of being rather tedious and taking a long time to come to the point. This is a poignant reminder of how deeply we have accepted his ideas and now take for granted this imaginative and fruitful approach. The heart of his idea is the beautifully logical combination of basic laws of chemistry with simple models of how the reactants come into contact with each other. The PFR and CSTR are readily imagined and widely applicable descriptions of how real large scale reactors actually behave. These models of the patterns of reactant contacting form the basis for much of present day chemical reaction engineering.

The reason for me reminding you of these foundations is to draw attention to the scarcity of further models beyond these two basic types. The bubbling gas fluidised bed is one of the very few additional models albeit a much more complicated one and necessarily more limited in application. This is why it is such a fascinating subject to chemical reaction engineers.

To return for a moment to the traditional models, they have been expanded, elaborated upon and mixed in all possible (and some impossible) proportions and, with the willing aid of modern computational methods, developed to a level of complexity that is in some cases out of touch with reality and certainly no longer physically imaginable but this process of paralysis by analysis is the fate of many originally simple ideas. Fortunately fluidised bed reactor models are still reasonably closely related to what actually happens and by and large the models can still be imagined as physical realities.

At this stage I should for the benefit of those less familiar with fluidised beds spend a few minutes describing the principal features that govern their behaviour as reactors (9). All powders bubble when fluidised by gas at velocities in excess of a minimum value and look rather as in Figure 1 which is a photograph of a two-dimensional bed sandwiched between glass plates. It is the bubbles that cause particle mixing and bring about the high heat transfer rates that can occur between the bed and walls or immersed surfaces. They are responsible for the high degree of uniformity, especially of temperature and particle composition, within the bed and without them the system would behave more or less as a packed bed.

Bubbling leads immediately to the concept of two phases with part of the gas flowing interstitially amongst closely spaced particles and the rest flowing in the form of bubbles. The interstitial flow remains constant at the minimum fluidisation value although fine powders may expand a little and permit a rather larger flow. The bubble flow is therefore easily evaluated from

$$Q_B = Q - Q_i$$

and the fact that $Q_i \doteq Q_{mf}$.

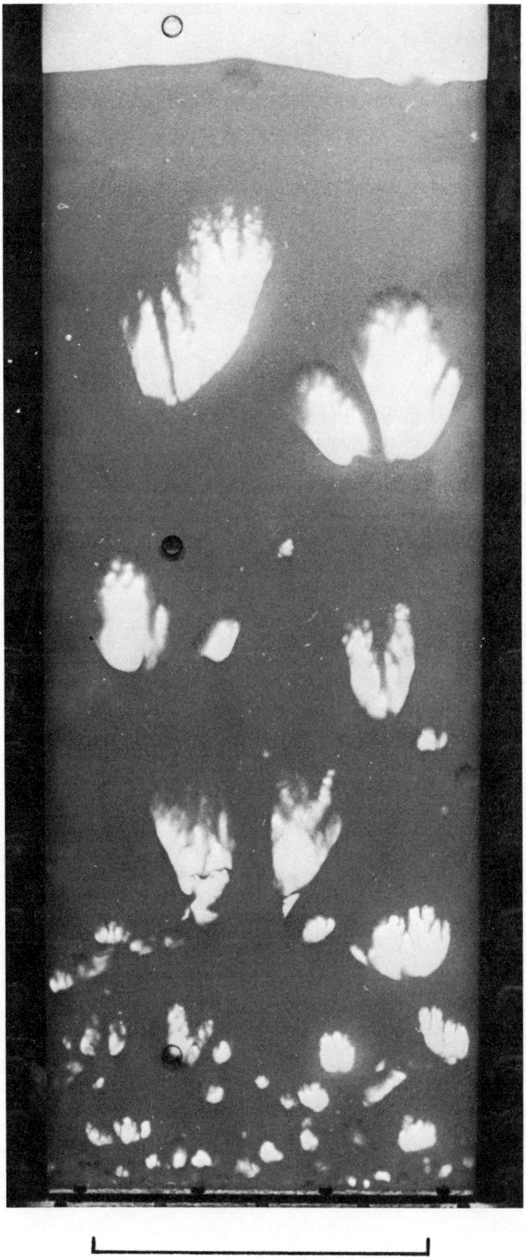

Figure 1. Bubbles in a two-dimensional fluidized bed.

The bubbles coalesce readily so that they may be small and numerous at the bottom and large and few at the top whilst maintaining substantially constant flow. Average bubble size increases with height as indicated in Figure 2. These simple facts allow an immediate consideration of reactor vessel sizing (10). Production requirements will determine the reactant gas feed rate and bed diameter fixes velocity and thence the bubbling rate. Heat transfer considerations and blower capacity are likely to set limits to the bed height or aspect ratio and bed diameter must accommodate the bubbles. What would be a bubble in a large reactor at a given height and velocity would be a slug in a narrow one as Figure 3 indicates. The slugging bed generally gives better gas/solid contacting but must be modelled differently from the bubbling bed and requires greater free-board. In the past pilot plants have operated in the slugging regime and reduced performance was experienced when scaling up simply by increasing diameter but no one should make that mistake today.

Particle mixing is caused by the bubbles, partly be shear displacement or drift but also by the bulk transport of particles in the bubble wake. Bubbles may also cause segregation if there are different kinds of particles present. Unlike other kinds of mixers, segregation is insensitive to particle size difference but particularly sensitive to density difference. In a binary system of particles segregation increases approximately as particle density ratio to the power 5/2 but with particle size ratio only to the power 1/5 (11). This can cause problems in, for example, coal combustion where char has a markedly lower density than ash and also in some ore reduction processes using coke.

Chemical reactor models invariably start from the two-phase theory (12). The interstitial flow is assumed to be in good and continuous contact with solids whilst some by-passing occurs in the bubble phase. There is, however, very little axial or radial mixing of the gas. There may be some exchange between the two phases and Figure 4 depicts this kind of model.

A major reason for by-passing in the bubble phase is the formation of clouds or spherical vortices centred on the bubbles, a situation that occurs when bubbles rise faster than interstitial gas which is commonly the case. Gas is obliged to flow upwards through the bubble because of pressure difference but as it re-enters the dense phase ahead it is caught in a stream of particles flowing downwards relative to the bubble and, when the flow is fast enough, it is trapped and dragged downwards until pressure difference pushes it back into the bubble. This is illustrated in Figure 5. Only that portion of the cloud or vortex that extends beyond the bubble is in contact with particles and therefore able to react. In many realistic situations this proportion can be very small. At any instant only this proportion of the total cloud is reacting so that the effective rate constant is reduced. The whole is circulating as a vortex with

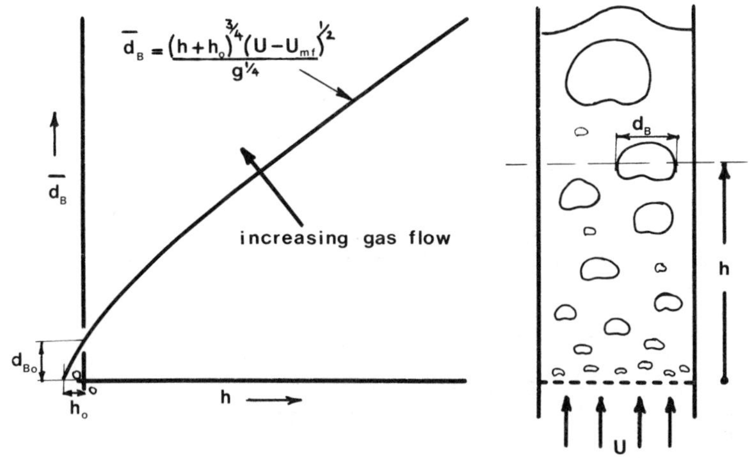

Figure 2. Bubbles size increases with bed height and flow rate.

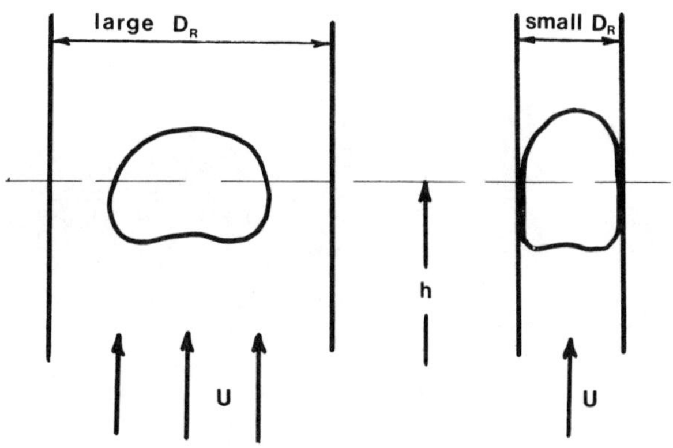

Figure 3. A bubble in a large bed becomes a slug in a small one.

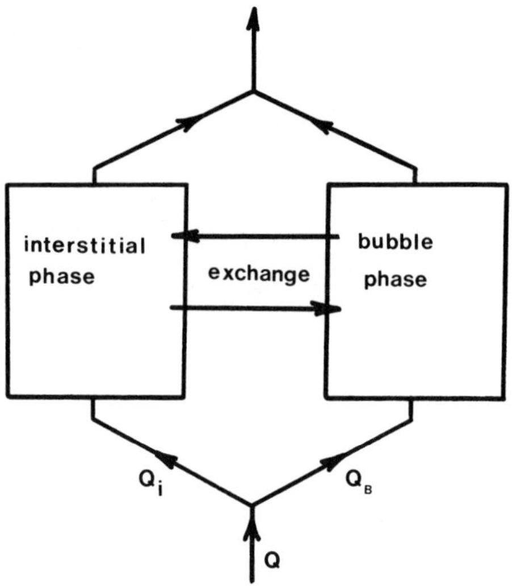

Figure 4. Basic two-phase bubble model.

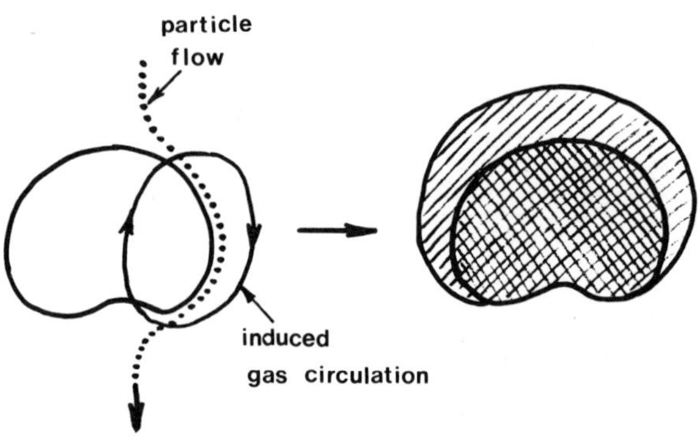

Figure 5. Particle flow induces gas circulation and so forms clouds.

upward velocity through the empty space of the bubble of similar magnitude to minimum fluidisation velocity. With fine particles this is small so that little circulation occurs during the residence time of a bubble in the bed. This is the very situation where the cloud cannot penetrate far into the dense phase so the overall effect is poor contacting of bubble phase gas with particles.

Severe by-passing in the bubble phase and very poor gas mixing can make the fluidised bed an extremely poor chemical reactor but at least the major reasons for this are now understood. The prospects of success for a given process can be judged at an early stage and design can minimize by-passing usually by choosing particle size appropriately.

What development and progress has occurred during the last ten years or so? Experience has accumulated and a number of fluidised bed processes are operated successfully. Industrial companies are reluctant to disclose too many details of a successful and productive reactor but I have seen beds as large as 17 m in diameter very uniformly and stably fluidised which could be shut down and re-started without trouble.

During the last decade much interest has centred on the behaviour of large particles (diameters in millimeters rather than microns) as required in fluidised coal combustion. The topic has generated two large conferences recently in the U.K. alone (13, 14) and many more in the U.S.A. This is the subject of the next Plenary Lecture so I will only mention some features briefly. Large particles are necessary in this application because high flow rates are required since the fluidising air is also combustion air and a high intensity of heat generation is aimed for. In order fully to convert oxygen to CO_2, coal concentration in the bed is low (a few percent) and most of it is ash or a chosen refractory granular material such as sand. Limestone may be additionally included to adsorb SO_2. These particles must be fairly large to avoid elutriation. Furthermore, the bed is shallow to minimise the pressure drop the fans are required to overcome. There is likely to be a lot of heat exchanger surface within the bed. Of course, gas velocity and hence the need for large particles can be reduced by increasing absolute pressure.

These conditions have not been widely studied in the laboratory because of the obvious difficulties of providing high air velocities (a few meters/sec) on a scale that is large compared with the size of the particles. Certainly interstitial gas flow will far exceed bubble rise velocity and there is no possibility of cloud formation as shown in Figure 5. Instead gas will flow straight through the bubbles and the overall pattern will approximate to plug flow. Because of the large difference in density between coal char and other bed material the former will tend to float. This is a complication imposed upon what would otherwise be a well mixed particle bed. Efficient in-bed combustion requires uniform dispersion of coal but the process of particle segregation is quite well understood (15) and the system can be designed to be reasonably uniform under operating conditions. Satisfactory models exist for prediction of combustor performance although there is some difficulty in correctly estimating the temperature of a burning coal particle. The major design problems are of the broadly mechanical kind such as coal feeding and initial distribution, heat exchanger design and ash removal.

At the other end of the spectrum there is continuing interest in the behaviour of fine particles (approximately less than 100 μm) because catalyst powders are generally in this range. Fluid bed cracking catalyst is the best known example. Geldart (16) made an important contribution when he proposed a classification of powders and labelled this fine material Type A. The characteristic feature of them is that they expand uniformly at velocities not greatly in excess of U_{mf} as illustrated in Figure 6. Only at some higher velocity, U_{mb} do they begin to bubble. Values of U_{mb}/U_{mf} can be as high as 3 or 4 and the bed volume can double in extreme cases before bubbling begins.

If the dense phase expands this immediately affects the reactor model because more gas will then flow via the favourable interstitial phase. Most models readily allow for this change given that the true division of flow can be predicted. Unhappily it has so far only been possible to measure this division experimentally at fairly low flow rates, well below those employed in commercial reactors. Certainly at velocities up to about 15 cm/s much more gas flows interstitially through Geldart type A powders than minimum fluidisation flow (17).

Industrial reactors generally operate at very high velocities (of order 1 m/s) much in excess of terminal falling velocity for at least the finest powder fractions. Powder is continually elutriated and returned to the bed via cyclones. Under these conditions there is disagreement as to whether or not bubbles retain their identity and such beds have been described as "turbulent" or "fast fluidised". What little evidence there is supports the continued existence of bubbles but now in a much disturbed or heterogeneous dense phase and with a less definite shape. Until more is known about this physical situation it is not easy to see how the bubbling bed reactor models should be modified correctly to describe this flow regime.

It is too simple to assume that average particle size is an adequate index of powder type in Geldart's classification. Certainly type A powders depend strongly on distribution of size and particularly on the proportion at the lowest end of the range. There is plenty of industrial experience to support this view and plant operators usually acknowledge the importance of maintaining a certain proportion of fines in the reactor. Whilst the vital role of the fines fraction has been recognised it is a difficult dependence to study sytematically.

The two-phase bubbling bed model is capable of many minor adjustments and has given numerous academics a lot of fun playing

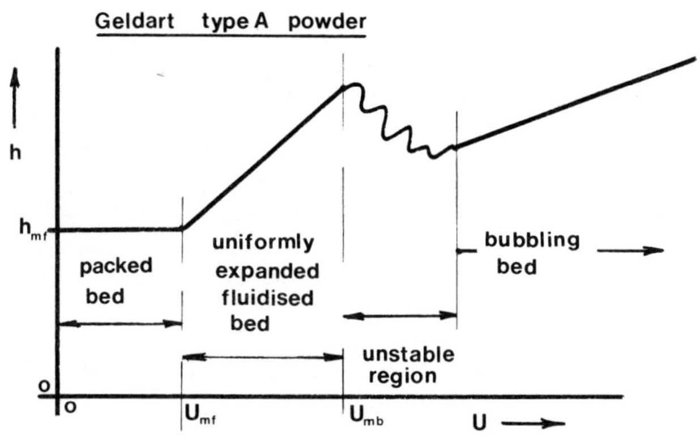

Figure 6. The expansion of type A powders.

variations on the basic tune. There are more models than data against which to test them and suitable data for this purpose are not easily obtained. Whatever the model chosen it is commonly observed that the reaction rate is unusually high in the bottom few centimeters of the bed and this is often referred to as the distributor effect.

The distributor plate in a full scale reactor is usually drilled with a number of holes several millimeters in diameter and the gas velocity through these is high (tens of m/s). It is easy to imagine that this produces a spout or jet of gas above which a stream of bubbles occurs. Powder could be entrained into the jet as in a spouted bed and it has been imagined that this potentially good gas/solids contacting arrangement is responsible for a high rate of chemical conversion (18).

Although attractive this model is quite wrong simply because gas does not form a jet above the distribution orifice but enters the bed in the form of bubbles just as, for example, air blown into water (19) - Figure 7. Here I must pause to emphasise the importance of identifying correctly the mechanism by which events occur which can only be done by suitable experimental observation.

The evidence comes from X-ray ciné photographs taken at 50 frames/s with an exposure time of 1 ms which conditions allow the very rapid events to be clearly followed. As Figure 8 shows, the bubble forms as a near perfect sphere above the hole, detaches and rises and within a very short distance, assumes its characteristic shape with an indented base as the wake forms. Essentially the same bubbling occurs if the hole is covered with a bubble cap, if gas enters downwards through a pipe buried in the bed or if gas enters horizontally through a hole in the wall. In this last case there is no observable lateral penetration and bubbles rise vertically from the orifice. This mode of entry has been observed at gas velocities up to 70 m/s and through holes as large as 16 mm diameter. All this has been seen with a variety of particles but not very large ones (d_p measured in mm) where there is reason to think behaviour may be different.

One reason for the persistence of a wrong concept of how gas enters is the fact that it does so differently in a two-dimensional fluidised bed, an arrangement chosen because of the ease of observation. In this case it is quite possible to establish a permanent jet stabilised by the walls. In a narrow slab bed particles can only flow towards the hole over a restricted region and cannot always move fast enough to close the hole periodically. There is no such difficulty when particles can approach within the full 360°. Possibly a jet could be established in a cylindrical bed by erecting suitable baffles to hinder particle flow but this would have to be a deliberate arrangement.

I would like to spend a few minutes over this detail because it leads to a simple and satisfying model for chemical reaction in the distributor zone. When an insoluble gas enters a liquid it is no surprise that bubbles form because of the incompatibility of the two phases. Furthermore bubble size and frequency must be equivalent to the gas flow rate. When gas enters a fluidised bed the phases are not incompatible because gas may form bubbles, flow interstitially or transfer between the two. It is a fact of observation that at the moment of detachment bubble size and frequency accounts for only about half the orifice gas flow. About two bubble diameters higher they have grown in size and now account for all the flow. Frequency is essentially independent of flow rate and constant at about 8/s.

Unlike formation in a liquid the boundary of a fluidised bed bubble can only expand by gas flowing across it to produce the drag force that will cause the particles to move appropriately. During the time that a bubble grows to the size shown in Figure 9 the gas that produced it has advanced to fill the volume indicated by the outer broken line. The annular region above and around the bubble now contains an excess of gas and so the powder voidage must increase. This is unstable and as the bubble detaches and rises through the expanded dense phase the powder relaxes and and returns the excess gas to the bubble. This appears to be completed by the time it has risen about one diameter (of order 1/10 second) and thereafter is of constant volume until it coalesces.

Considering now the consequences for chemical reaction, much of the reactant gas that ultimately forms a bubble will first enter the interstitial phase and enjoy a brief moment of intimate contact with the particles. Later it may form a cloud as in Figure 5 with limited access to particles. Hence the high reaction rates associated with the process of bubble formation and limited to the bottom layer of the bed of thickness up to twice the initial bubble diameter - a few centimeters at most.

As yet there is no fluid dynamic model that describes in quantitative detail the bubble formation process but it is barely necessary for a reaction engineering model. It is adequate to assume that entering reactant gas passes in plug flow through the bottom layer of particles, say, one initial bubble diameter deep and thereafter forms bubbles. Initial bubble diameter is readily estimated from the known flow through the orifice and the fact that frequency is about 8/s. Above this distributor layer the two-phase bubble model can be applied.

The other end condition where the bubbling bed model is inappropriate is above the bed where there may be reaction in the free board region. With fine powders where there is appreciable elutriation gas and particles may remain in contact with further opportunity for reaction. This situation has not attracted the attention of many modellers but at least one model predicts that considerable reaction can continue under certain circumstances (20).

The last area I wish to mention is the effect on fluidisation of changing temperature and pressure. Not very much funda-

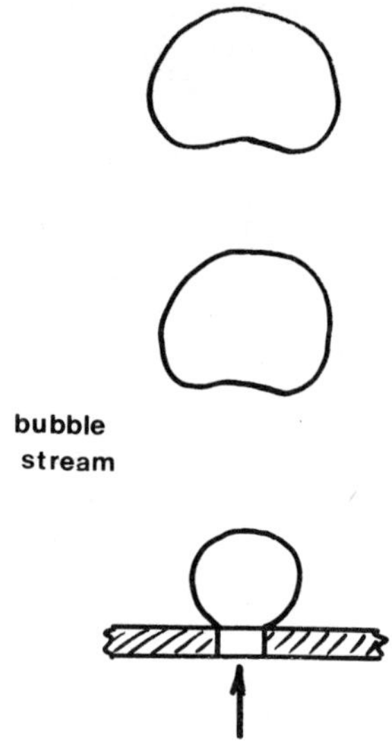

Figure 7. Gas enters the bed as a stream of bubbles.

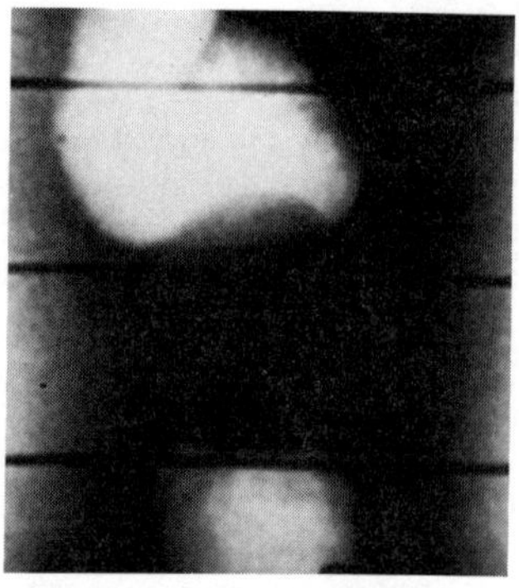

Figure 8. x-Ray photograph of bubble formation at an orifice.

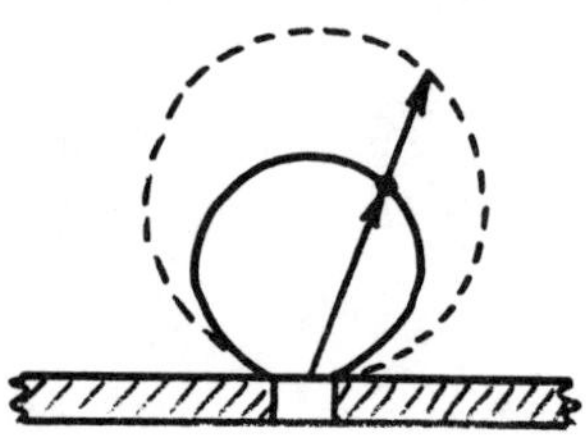

Figure 9. Gas flow causes the bubble boundary to expand.

mental work has been reported on the former (21, 22) but there is increasing interest in the latter (23-29). It is becoming fairly clear that when compared at the same superficial gas velocity, bubble size decreases as pressure increases. The reason for this seems to be that proportionately more gas flows interstitially as pressure increases and bubbles are smaller because the reduced flow gives less opportunity for growth by coalescence. Powders that normally behave as Geldart type B (i.e. $U_{mf} = U_{mb}$) become type A at quite modest increases in pressure. This is good news from a chemical reaction engineering point of view and the bubbling bed models continue to apply and only require the changed division of flow to be quantified.

At pressures greater than about 80 bar some interesting hydrodynamic changes occur. Bubbles begin to lose their identity and at quite low gas velocities the bed takes on the general appearance of a "turbulent" or "fast fluidised bed". This again is advantageous for reaction engineering but few processes will be required to operate at such high pressures.

I turn finally to consider the direction future basic research should take. It is fundamental to the state of fluidisation that particles are supported by the drag of flowing gas. This force depends not only on the gas properties and velocity but also on the particle spacing and arrangement. In spite of the interest of fluid dynamicists little is known about the relationship between permeability, voidage and the flow conditions and yet it is this that decides the division of gas between the phases with important consequences for chemical reaction. It is evident that particle size and size distribution are factors determining permeability and that "fines" are important in this respect but it is difficult to understand why change of absolute pressure should change voidage and permeability. Understanding these things could greatly improve our ability to engineer fluidised bed chemical reactors and this should be a major object of basic research.

However deep our knowledge of the mechanism of fluidisation it is salutary to pause and consider the freedom of choice a chemical reaction engineer will have. Leaving aside details of mechanical design such as the distributor, baffles, heat exchangers and materials of construction, the variables are very limited. Gas density and viscosity and particle density will be determined by chemistry and thermodynamics and the only major variables remaining are gas velocity, bed height and particle size and size distribution. The first two are largely fixed by production rate and simple engineering considerations and particle size is about the only thing left to choose. If we knew more about how to make this choice our designs might be much improved.

List of Symbols

d_B	bubble diameter
D_R	reactor diameter
h	bed height
Q	volumetric flow rate into the bed
Q_B	bubble flow rate
Q_i	interstitial flow rate
Q_{mf}	minimum fluidisation flow rate
U_{mb}	minimum bubbling flow velocity
U_{mf}	minimum fluidisation flow velocity

Literature Cited

1. Pyle, D.L. Advances in Chemistry Series 109, 1972, 106.
2. Rowe, P.N. Proc. 2nd ISCRE; Elsevier: Amsterdam, 1972, A9.
3. van Swaaij, W.P.M. Advances in Chemistry Series 72, 1978, 193.
4. Rudolph, P.F.H. Proc. 4th ISCRE 2; Dechema: Frankfurt, 1976, 537.
5. Kunii, D. Chem. Eng. Science 1980, 35, 1887.
6. Levenspiel, O. Chem. Eng. Science 1980, 35, 1821.
7. Denbigh, K.G. Trans. Faraday Soc. 1944, 40, 352.
8. Danckwerts, P.V. Chem. Eng. Science 1953, 2, 1.
9. Davidson, J.F.; Harrison, D. (Editor) "Fluidisation"; Academic Press: New York, 1971, 121 et seq.
10. Rowe, P.N. Chemistry & Industry No. 12 1978, 424.
11. Rowe, P.N.: Nienow, A.W.; Agbim, A.J. Trans. I. Chem. E 1972, 50, 324.
12. Kunii, D.; Levenspiel, O. "Fluidisation Engineering"; John Wiley: New York, 1969.
13. "Fluidised Combustion" 1975, London: Institute of Fuel, Symp. Series No.1.
14. "Fluidised Combustion - systems and applications" 1980, London: Inst. of Energy, Symp. Series No.4.
15. Nienow, A.W.; Rowe, P.N.; Cheung, L.Y.-L. Powder Tech'y 1978, 89.
16. Geldart, D. Powder Tech'y 1973, 7, 285.
17. Rowe, P.N.; Yates, J.G.; Santoro, L. Chem. Eng. Science 1978, 133.

18. Zenz, F.A. "Fluidisation"; Pirie, J.M., Ed.; I. Chem. E. Symp. Series No.30, 1968, 136.
19. Rowe, P.N.; MacGillivray, H.J.; Cheesman, D.J. Trans. I. Chem. E. 1979, 57, 194.
20. Yates, J.G.; Rowe, P.N. Trans. I. Chem. E. 1977, 55, 137.
21. Botterill, J.S.M.; Yeoman, Y. "Fluidization"; Grace, J.R.; Matsen, J.M., Eds.; Plenum Press: New York, 1980, 93.
22. Desai, A.; Kikukawa, H.; Pulsifer, A.H. Powder Tech'y 1977, 16, 143.
23. Harrison, D.; Davidson, J.F.; de Kock, J.W. Trans. I. Chem. E. 1961, 39, 201.
24. Clift, R.; Grace, J.R.; Weber, M.E. Ind. Eng. Chem. Fund. 1974, 135, 45.
25. Carvalho, J.R.F. Guedes de; Harrison, D. Inst. of Fuel Symp. Series No.1; 1975, B1.
26. Varadi, T.; Grace, J.R. "Fluidisation"; C.U.P.: 1978, 55.
27. Carvalho, J.R.F. Guedes de; King, D.F.; Harrison, D.: ibid. 59.
28. Subzwari, M.P.; Clift, R.; Pyle, D.L.: ibid. 50.
29. Rowe, P.N.; MacGillivray, H.J. Inst. of Energy Symp. Series 4, 1980, IV, 1.

RECEIVED April 27, 1983

4

Fluidized-Bed Coal Combustion: Controlling Parameters

ADEL F. SAROFIM

Massachusetts Institute of Technology, Department of Chemical Engineering, Cambridge, MA 02139

The factors controlling the performance of atmospheric-pressure fluidized-bed combustors (AFBC's) are illustrated by order-of-magnitude calculations of the design and operating variables for a hypothetical 100 MW_t boiler. Bed height, bed cross-sectional area, immersed heat-transfer surface area, sorbent feed size and rate, bed temperature, fluidizing velocity and coal feed size are selected by consideration of the fluids mechanics, heat transfer, and kinetics of the governing gas-solid reactions. The fluid mechanics and heat transfer in AFBC's are shown to be differ distinctly from those in more traditional fluidized bed reactors.

Fluidized bed combustion provides one option for utilizing some of our large coal reserves for purposes of energy generation. Its advantages over alternative methods of direct coal utilization, e.g., suspension firing of pulverized coal and the burning of lump coal on stokers, is that it is able to handle coals of widely varying composition, including low-grade coals with high ash content, and it provides a capability of in-bed sulfur capture.

The development of the technology of AFBC's has been pursued vigorously since the early sixties. Small (<3MW) units are in commercial service in both the United Kingdom (1) and the People's Republic of China, with over 200 units in use in the latter, some with over 40,000 hours of operational experience (2). Larger demonstration units (>20MW) have been constructed and operated (1, 2, 3). At present several firms offer designs of AFBC's for industrial steam generation, and the development of atmospheric-pressure and pressurized fluidized-bed combustors for central-station electric-power generation is being pursued at a pilot-scale level (4, 5). Consideration in this paper will be restricted to units operated at atmospheric pressure (AFBC's) since these are of greater current commercial interest.

0097–6156/83/0226–0065$10.00/0

Fluidized-bed combustors are complex chemical reactors, a fact often overlooked because their primary purpose is to generate steam and electricity. The reactions of importance in fluidized beds include those relating to coal combustion,

$$\text{coal} \rightarrow \text{char} + \text{volatiles (including } NH_3 + H_2S)$$

$$\left.\begin{array}{l} \text{volatile} + O_2 \rightarrow CO_2 + H_2O \\ 2C + O_2 \rightarrow 2CO \\ C + CO_2 \rightarrow 2CO \\ CO + 1/2\ O_2 \rightarrow CO_2 \end{array}\right\} + NO + SO_2$$

to sulfur capture,

$$\begin{array}{l} CaCO_3 \rightarrow CaO + CO_2 \\ CaO + SO_2 \rightarrow CaSO_3 \\ CaSO_3 + 1/2\ O_2 \rightarrow CaSO_4 \\ SO_2 + 1/2\ O_2 \rightarrow SO_3 \\ CaSO_3 + 1/2\ O_2 \rightarrow CaSO_4 \end{array}$$

and to nitric oxide formation and destruction,

$$\begin{array}{l} 2NH_3 + 3/2\ O_2 \rightarrow 2NO + 3/2\ H_2O \\ NH_3 + NO + 1/4\ O_2 \rightarrow N_2 + 3/2\ H_2O \\ NO + C \rightarrow 1/2\ N_2 + CO \\ NO + CO \xrightarrow[\text{solid}]{} 1/2\ N_2 + CO_2 \\ NO + CaSO_3 \rightarrow CaSO_4 + 1/2\ N_2 \end{array}$$

From this partial listing one can find reactions belonging to many of the important classes of homogeneous and heterogeneous reactions.

The performance of a fluidized bed combustor is strongly influenced by the fluid mechanics and heat transfer in the bed, consideration of which must be part of any attempt to realistically model bed performance. The fluid mechanics and heat transfer in an AFBC must, however, be distinguished from those in fluidized catalytic reactors such as fluidized catalytic crackers (FCCs) because the particle size in an AFBC, typically about 1 mm in diameter, is more than an order of magnitude larger than that utilized in FCC's, typically about 50 μm. The consequences of this difference in particle size is illustrated in Table 1. Particle Reynolds number in an FCC is much smaller than unity so that viscous forces dominate whereas for an AFBC the particle Reynolds number is of order unity and the effect of inertial forces become noticeable. Minimum velocity of fluidization (u_{mf}) in an FCC is so low that the bubble-rise velocity exceeds the gas velocity in the dense phase (u_{mf}/ε_{mf}) over a bed's depth; the FCC's operate in the so-called fast bubble regime to be elaborated on later. By contrast the bubble-rise velocity in an AFBC may be slower or faster than the gas-phase velocity in the emulsion

Table I.

Differences in Fluidization and Heat Transfer of AFBC's and Fluidized Catalytic Reactors (e.g. FCC's)

	FCC	AFBC
d_p	$\sim 50 \mu m$	~ 1 mm
u_{mf} (m/s)	$\sim 10^{-3}$	~ 0.5
Re_p	$\sim 10^{-4}$	~ 3
u_{br}	$\gg u_{mf}/\varepsilon_{mf}$	$\simeq u_{mf}/\varepsilon_{mf}$
$\tau_{particle(s)\ heating}$	$\sim 10^{-2}$	~ 1

phase, leading to operation in either the slow or fast bubble regime. Finally the characteristic heating time of particles in an FCC is small relative to the residence time of solids near a tube surface and the heat transfer rate is then controlled by the penetration of a thermal waves into the solids as first proposed by Mickley and Fairbanks (6). For an AFBC, the characteristic heating times are long compared to the particle residence time at the wall and the particles can be treated as being at the bulk temperature for purposes of heat transfer calculations (7).

The paper will consider the parameters controlling the design and operation of an AFBC, shown schematically in Figure 1. The questions to be addressed are what are the factors that guide the selection of bed height, freeboard height, coal feed location, immersed heat-transfer surfaces, sorbent size and feed rate, coal feed size, bed temperature and fluidizing velocity. In order to help focus the discussion, order-of-magnitude calculations will be performed for the conceptual design of a boiler designed to produce 100 MW of steam at 644 K, fired with a coal of elemental composition $CH_{0.7}\ S_{0.02}\ N_{0.03}\ O_{0.04}\ (H_2O)_{0.07}$, a heating value of 29,100 kJ/kg, and ten percent ash by weight. The boiler will be operated with 20 percent excess air.

Pressure Drop Across Bed (Bed Height Selection)

The bed height is determined by consideration of the work required to overcome the pressure drop across the bed and distributor plate. The pressure drop across the bed is simply the weight of the bed.

$$\Delta P = \rho_s g(1-\varepsilon)h \qquad (1)$$

where ε and h are the void fraction and height of the unexpanded bed, and ρ_s is the density of the bed solids. The additional pressure drop across the distributor plate will here be taken to be one third that across the bed. The fan work, and air flow rate

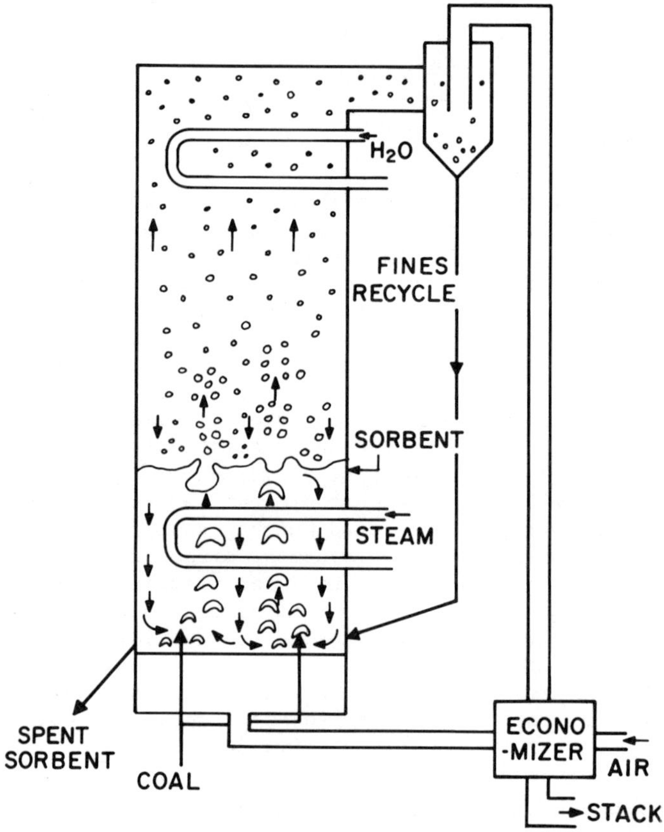

Figure 1. Schematic of atmospheric pressure fluidized bed combustor.

is the product of the volumetric air flow rate and the pressure drop, may be expressed as an equivalent amount of fuel $\Delta\dot{m}_F$, given the volumetric air requirement v_a per mass of fuel. The fraction of the fuel flow rate $\dot{m}_F$ needed to supply the energy to drive the fans is readily shown to be

$$\frac{\Delta\dot{m}_F}{\dot{m}_F} = \frac{4}{3}\,\frac{v_a\rho_s(1-\varepsilon)gh}{(-\Delta H_C)\eta_e\eta_f} \tag{2}$$

where $-\Delta H_C$ is the heat of combustion, η_f the fan efficiency, and η_e the efficiency of conversion of thermal to electrical energy. For realistic values of the coefficients: ρ_s = 2400 Kg/m^3, ε = 0.5, η_e = 0.3, η_f = 0.6, v_a = 24.5 m^3/Kg coal,

$$\frac{\Delta\dot{m}_F}{\dot{m}_F} = 0.028\ h \tag{3}$$

where the bed height h is in meters. The unexpanded bed height needs to be kept low in order to minimize pumping energy losses, and a value of about 0.7 m is often selected to keep the energy loss for fan power to 2 percent of the fuel's heating value. A consequence of this constraint on maximum bed height is that fluidized bed cross-sectional area is proportional to the capacity of a boiler, resulting in bed widths much larger than bed depth for larger units; this can pose problems of ensuring uniform fuel distribution over the bed, as will be discussed later. Inasmuch as the fan power loss is proportional to the volumetric air requirement v_A per unit mass of fuel (Eq. 2), the bed height corresponding to a given $\Delta\dot{m}_F/\dot{m}_F$ is proportional to pressure so that pressurized FBC's may be operated with deep beds and much smaller cross-sectional areas than AFBC's.

Sulfur Capture (Selection of Bed Temperature and Particle Size)

The sulfur sorbent that has received the most attention is limestone because of its widespread availability and low cost. Unfortunately the conversion of limestone to calcium sulfate results in a volumetric increase, and it can be readily shown (8) that for particles that do not permit volumetric expansion on reaction the maximum conversion attainable is about 59 percent (the value is dependent on the specific volumes of reactant and product). This constraint can be avoided by use of dolomite, since the MgO is not sulfated and the void volume produced by the decomposition of $MgCO_3$ and $CaCO_3$ is sufficient to permit complete calcium utilization, but at the expense of an added weight of sorbent and added energy requirement for calcination. The above conclusions follow from consideration of the stoichiometric relations:

$$CaCO_3 + SO_2 + 1/2\ O_2 \rightarrow CaSO_4 + CO_2 \qquad \Delta H = -325\ kJ/mole \qquad (4)$$

$$CaCO_3.MgCO_3 + SO_2 + 1/2\ O_2 \rightarrow MgO.CaSO_4 + 2CO_2$$

$$\Delta H = -210\ kJ/mole \qquad (5)$$

The rate constant for sulfation of a limestone is shown as a function of the fractional sulfation m of calcium, for several particles sizes, in Figure 2, taken from the measurements of Fieldes and Davidson (9). The rate of sulfation is seen to decrease with extent of sulfation approaching zero at a maximum sulfation m_∞. Both the rate constant K_s at a given sulfation and the maximum sulfation m_∞ decrease with increasing particle size; the data may be correlated by one curve when plotted versus m/m_∞, a correlation which has proved useful in developing models of sulfur capture for AFBC's (10).

The maximum sulfation observed is considerably smaller than the theoretical limit of 59 percent. This observation, together with the decrease in rate constant K_s with extent of sulfation, can be explained by models which assume that the reaction is retarded by a product layer of $CaSO_4$ of low porosity through which the SO_2 must diffuse in order to reach the unreacted CaO. Such a hypothesis has formed the basis for the development of several models to quantitatively fit sulfation data, of which the grain model (11, 12) and the pore-plugging model (13, 14) are the most notable.

In the grain model, it is assumed that the CaO consists of spherical grains of uniform size distributed in a porous matrix. The rate of reaction is controlled by the diffusion of SO_2 through the porous matrix and the product $CaSO_4$ layer formed on each grain (11). Allowance can be made for a finite rate of the CaO/SO_2 reaction (12). The models have been found to describe experimental data for many limestones (13) by adjusting the constants in the model, most notably the diffusivity through the product layer.

In the pore-plugging model the calcined limestone is assumed to consist of calcium oxide perforated by cylindrical holes of uniform diameter. The SO_2 diffuses through the pore and the product $CaSO_4$ to unreacted CaO. Because of the increase in volume on reaction, the pore diameter decreases with a maximum decrease at the opening where the diffusion path is shortest. As a consequence the reaction rate falls to very low values as the pore mouth plugs. The model has been developed without (14) and with (15) allowance for a kinetic resistance due to the CaO/SO_2 reaction. As with the grain model, the pore-mouth plugging model can adequately describe the decrease in reaction rate with increasing extent of reaction and with decreasing particle size by appropriate adjustment of the diffisuvity through the product layer.

The reaction of SO_2 with calcined limestone is found to be first order with respect to SO_2 but the dependence on O_2 concentration is more complicated. Fieldes et al. (16) have observed

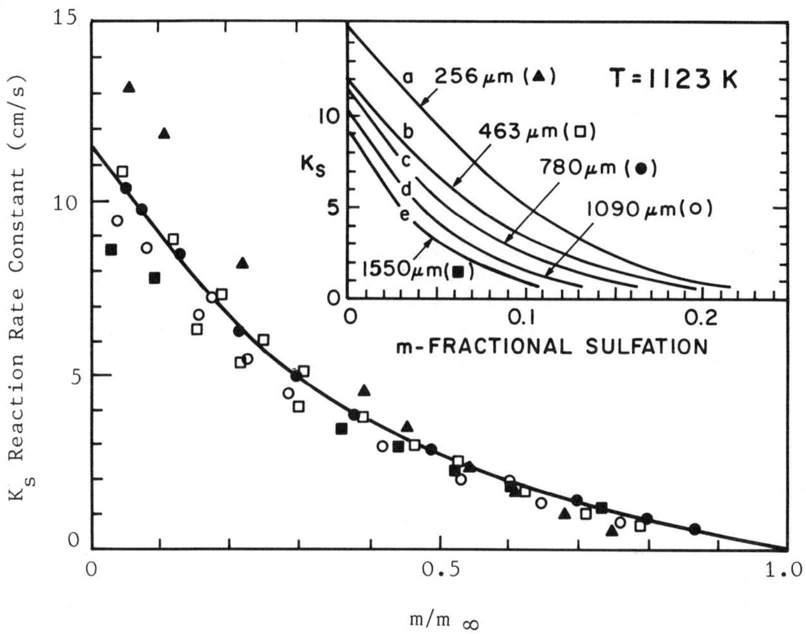

Figure 2. Rate constant for the reaction of SO_2 with calcined limestone as a function of m, the fractional sulfation, and m/m_∞, the fractional sulfation normalized with its asymptotic value (9).

that the maximum sulfation m_∞ of a variety of limestones decreased with increasing oxygen concentration. They attributed this decrease to an increase in SO_3 concentration with increasing oxygen; since the rate of reaction of SO_3 with CaO is high the SO_3 will react preferentially at the mouth of a pore and lead to the pore plugging.

The maximum sulfation m_∞ is expected to pass through a peak as temperature is increased since, for the levels of SO_2 and O_2 found in combustion products, $CaSO_4$ decomposes at temperature in excess of 1500 K. The maximum observed, however, is at lower temperatures. The maximum sulfation m_∞ is reported as occuring around 1120 K (17, 18). This provides the basis for the current selection of optimum bed temperature. The explanation for the peaking in sulfur capture at temperatures below that at which $CaSO_4$ decomposes has not been definitely established. A plausible argument for the decrease in reactivity with increasing temperature is that sintering of grains is accelerated at higher temperatures and that the maximum sulfation therefore decreases as a consequence of the increased grain size. Examples of the increase in grain size (19) and the decrease in maximum sulfation (9) with increasing temperature are shown in Figure 3.

In order to predict emissions from AFBC's it is necessary to couple a model of the sulfur capture of individual particles into a system's model which takes into account the SO_2 formation, removal, and transport. Because the single particle behavior is so complex, most such models (10, 20, 21, 22) use simplified, usually empirical, fits of single particle behavior determined, for example, from thermogravimetric analysis.

The optimum conditions needed to achieve the desired sulfur capture efficiency (90 percent) is dependent on sorbent type, coal type, and fluidized bed design. Representative conditions are a bed temperature of 1116 K, a sorbent particle of 1 mm, and a Ca/S ratio of about 3. The bed temperature selection is based on the admittedly shaky evidence that this is where the maximum sulfation peaks. The particle size selection is a compromise between a desire to use high gas velocities and the observed decrease in sulfation with increasing size (Figure 2). The Ca/S ratio corresponds to a maximum sulfation of about 0.3. For these conditions the mass feed rate of a limestone sorbent is 0.345 that of the coal feed and the flow rate of the partially sulfated spent sorbent is 0.285 the coal feed rate.

Clearly there is an incentive for reducing the amount of sorbent. Promising areas for research are on reducing grain size by minimizing grain growth (23, 24), increasing limestone utilization through partial hydration (25) and the development of regenerative sorbents (26).

Fluid Mechanics (Determination of Bed Cross-sectional Area and Flow Regime

The fluidizing velocity in a bed is bounded by the minimum fluidizing velocity u_{mf} and the terminal velocity u_t (27). The

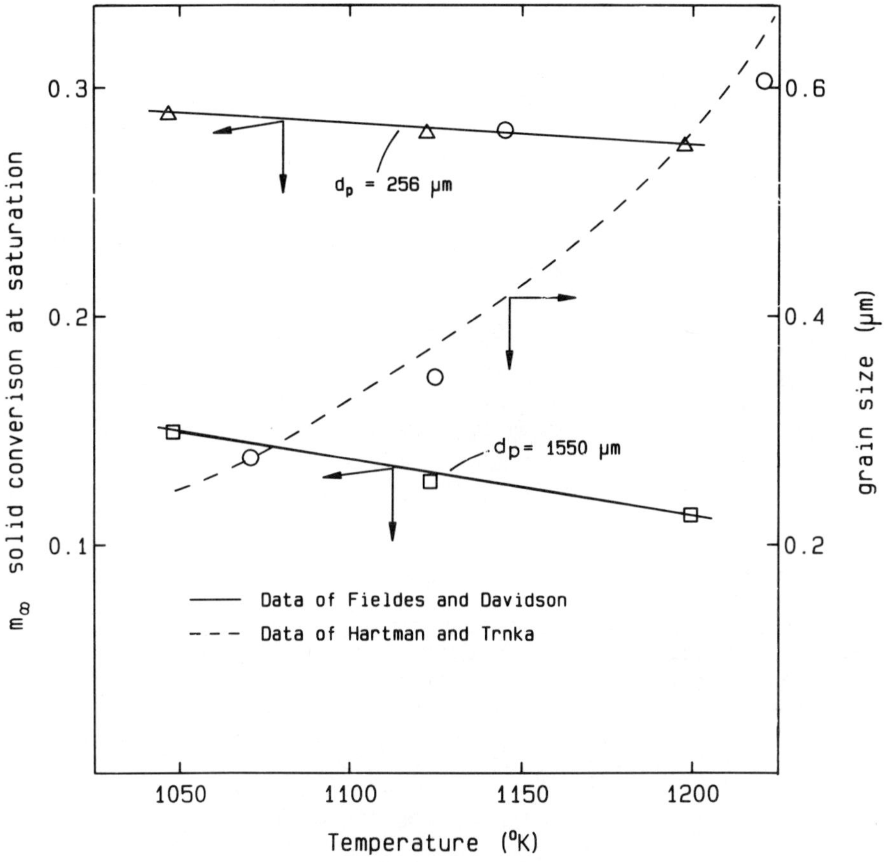

Figure 3. Effect of bed temperature on grain size and asymptotic fractional sulfation of limestone. Data from Refs. 9 and 19.

correlations for the bounding curves are given in Figure 4 where the dimensional coordinates are for the conditions in an AFBC (using a sphericity factor $\phi = 0.6$). At low velocities the slopes approach two as would be expected for viscous flow, and at the higher velocities at which inertial forces are dominant, the slopes approach 0.5. For FBC's, the particle sizes are about 1.0 mm. From Figure 4, this size is found to correspond to a minimum fluidizing velocity of 0.4m/s, the approximate transition point to the regime in which inertial forces begin to become appreciable.

From a viewpoint of minimizing the cross-sectional area of a bed, it is desirable to operate at as high a velocity above the minimum fluidizing velocity as is practical. The upper limit on fluidizing velocity is imposed in part by the increase in amount of elutriated material with increasing velocity. Elutriation rates increase approximately in proportion to the exponential of the ratio u_o/u_t of the fluidizing to terminal velocity, according to one of the available empirical correlations (28). From consideration of the trade-offs between decreased bed cross-section and increased elutriation rates a velocity of about 2 m/s is often selected as the nominal operating velocity in AFBC's. At this velocity, particles with diameters below 350 μm once entrained from the bed will be carried over to the particulate collection devices since their terminal velocities will be exceeded by the fluidizing velocity, as may be seen from Figure 4.

Once the fluidizing velocity is selected, the bed cross-section area is fixed, being given by the volumetric flowrate required for combustion $\dot{m}_F v_a$ divided by the fluidizing velocity u_o. For our test case of 100 MW thermal output, the cross-sectional area of the bed is 87 m^2. The approximate cross-sectional area of AFBC's may be estimated by noting that the velocity constraint of 2 m/s corresponds approximately to 1.1 MW thermal output per square meter of bed area.

The bubbling behavior of a bed is also determined by the fluidizing velocity u_o. According to the two-phase theory of fluidization (29), the excess velocity over that required for minimum fluidization passes through the bed in bubbles, which provides a useful, although oversimplified (30, 31) view of bubble flow.

The gas flow through the bubbles depends on whether the bubble rise velocity, $0.71\sqrt{gd_b}$ (29) is slow or fast relative to the gas velocity in the emulsion phase (u_{mf}/ε_{mf}). When the bubble rise velocity exceeds the emulsion velocity, the bubbles are said to be fast or clouded and the gas circulates through a cloud surrounding the bubble as shown schematically in Figure 5 (adapted from Ref. 32). When the emulsion velocity is fast relative to the bubble rise velocity the bubbles are said to be slow or cloudless and the emulsion gas uses the bubble as a by-pass (see Figure 5). The transition from slow to fast bubble depends upon the

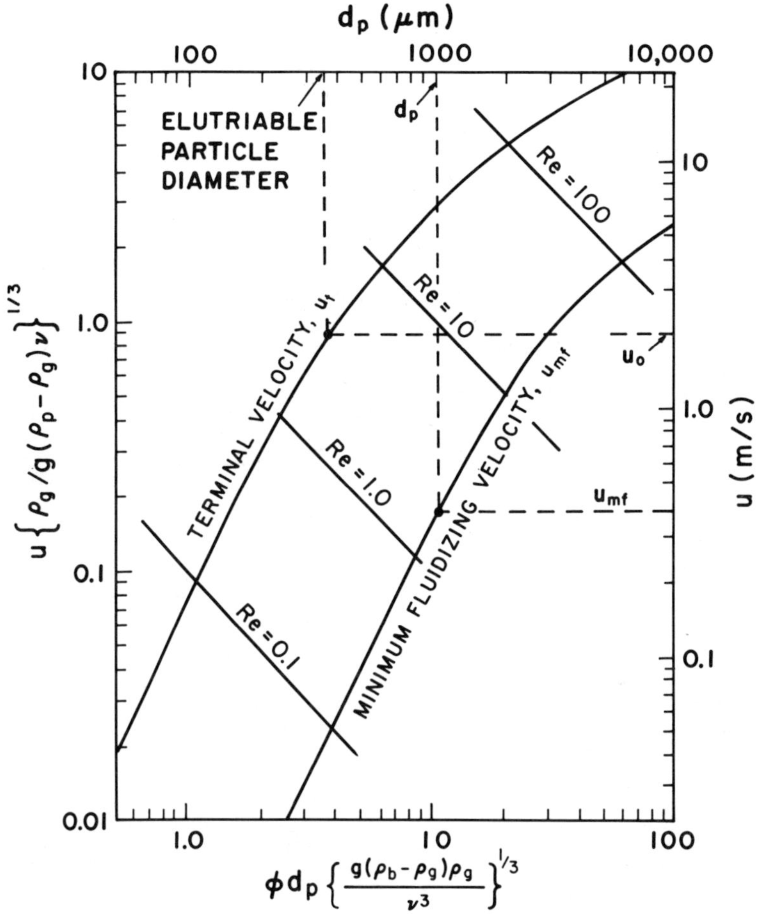

Figure 4. Correlations for the minimum fluidizing velocity u_{mf} and particle terminal velocity u_t as a function of particle size.

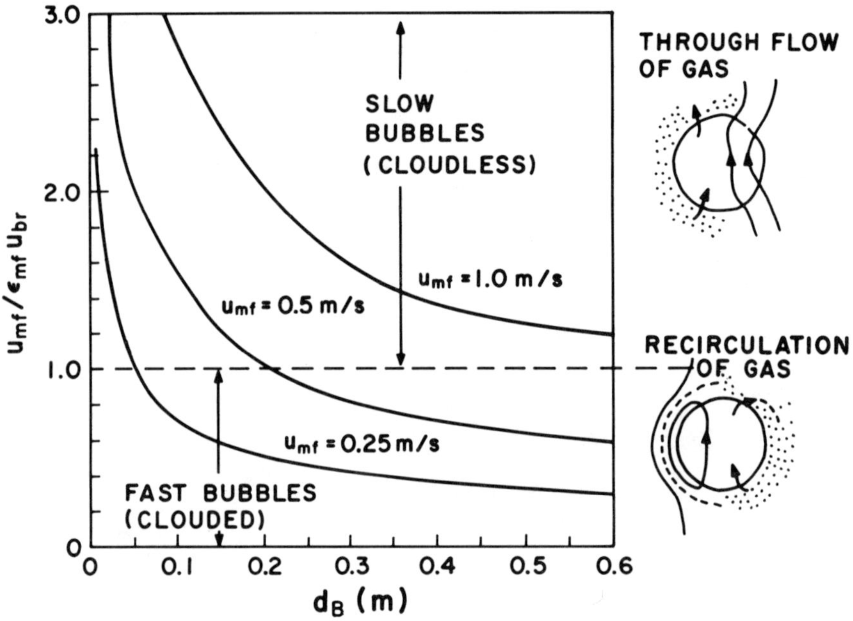

Figure 5. The dependence of the transition from the slow to fast bubble regime on bubble diameter d_B and minimum fluidization velocity u_{mf}.

minimum fluidizing velocity and the bubble diameter d_b, which may be estimated from the simple relation developed from the theory for bubble growth due to coalescence by Darton et al. (33).

$$d_b = 0.54\ (u_o - u_{mf})^{2/5}\ (h + 4\sqrt{A_o})^{4/5}/g^{1/5} \tag{6}$$

where A_o is the area of distributor plate per orifice and h is the height of the bubble above the plate. For u_o = 2 m/s and u_{mf} = 0.4 m/s and a bubble-cap plate with A_o = 2.5 x $10^{-2}m^2$, the bubbles increase in diameter from a value of 0.079 m at the distributor plate to a value of 1 m at the top of the bed (1 m). The bubble diameter at which the bubble velocity, $0.71\sqrt{gd_b}$, equals the emulsion velocity, u_{mf}/ε_{mf}, is 0.13m, and is the value encountered at a height of 0.1m above the distribution plate. The bubbles are, therefore, expected to start in the slow or cloudless bubble regime but accelerate into the free bubble regime at an elevation of 0.1m. As the bed particle diameter increases, u_{mf} increases and the fraction of the bed that is in the slow bubble mode will increase. Whereas FCC's operate primarily in the fast bubble regime AFBC's operate in the transition of the fast to slow bubble regimes.

The bubble motion will also determine the solid circulation in the bed, as a consequence of the entrainment of solids in the bubble wake. For an assumed ratio of wake to bubble volume, the volumetric circulation rate of solids per unit cross-sectional area of bed is equal to $(u_o - u_{mf})\alpha$. The solid circulation time is then equal to $h/(u_o - u_{mf})\alpha$. For the values u_o = 2m/s, u_{mf} = 0.4m/s, h = 1m, and a reasonable value for α of 0.1 to 0.2 (34), the solid circulation time is found to be about 3 to 6 seconds which is slow enough compared to limestone reaction times (see above) or char burnout times (see below) that the bed solids can be treated as being well-mixed.

Whether the gas can be treated as being well-mixed or in plug flow depends upon the velocity of the downward flowing solids relative to the minimum fluidizing velocity (34). If we neglect temporarily the cross-section of the bed occupied by the bubbles and the associated solids moving upwards, the velocity of the solids flowing downwards is then determined by their volumetric flow rate per unit cross-sectional area or $\alpha(u_o - u_{mf})$. When this velocity exceeds the instertitial gas velocity of u_{mf}/ε_{mf}, the gas is recirculated with the solids. The simple criteria for gas recirculation, developed in a more general form by Kunii and Levenspiel (34), is then

$$\frac{u_o}{u_{mf}} > 1 + \frac{1}{\alpha\varepsilon_{mf}} \tag{7}$$

For values of α of 0.1 to 0.2 and ε_{mf} = 0.5 we find that the transition from a plug flow to well-mixed gas occurs at a value of

u_o/u_{mf} in the range of 10 to 20, and once again we find a difference between the regime of operation of an FCC which operate with u_o/u_{mf} >20 and AFBC' which typically operate with u_o/u_{mf} below the transition point.

An other important effect of bubbles is to cause the bed to expand. The bed expansion is a consequence of the bubble hold up in the bed and can be calculated from the bubble flow rate and the bubble rise velocity (29, 34). For the considions here, a bubble void fraction of 0.3 is assumed, corresponding to an expanded bed height of about 1m. From the expanded bed height, bed cross-sectional area, and sorbent feed rate, it can be readily shown that the mean residence time of the limestone in the bed is about 20 hours.

Heat Transfer (Selection of Tube Area)

The heat transfer to the immersed tubes is by a combination of radiation, convection from gases flowing past the tube, and transfer from particles in the vicinity of the tube surface. Since the particle residence time near the tube walls is small relative to their thermal relaxation times (see Table 1), the temperature of the particles near the wall does not depart appreciably from the mean bed temperature. The heat transfer coefficient from the particles is then governed by conduction across the gap separating the particles from the wall. The sum of the conductive and convective fluxes can be calculated from the expression derived by Decker and Glicksman (7):

$$\frac{hd_p}{k} = [12 + 0.05\ (1 + 2\delta)(\frac{u_{mf}}{\nu} d_p)Pr] \tag{8}$$

where δ is the bed void volume. The above expression is for a tube surface in the emulsion phase, with the first term representing the conduction from the particles and the second the augmentation of the flux by convection. The radiative heat transfer contribution can be adequately approximated by the flux between two infinite parallel gray plates

$$h_r(T_b - T_w) = \frac{\sigma(T_b^4 - T_w^4)}{1/\varepsilon_b + 1/\varepsilon_w - 1} \tag{9}$$

where ε_w is the tube wall emmisivity, and ε_b is the effective emmissivity of a plane formed by the spherical particles. The contributions to the heat transfer by the above mechanisms is shown in Figure 6. As particle diameter increases the conduction contribution given by the first term in Eq. 8 falls rapidly but compensation is provided by the increased convection contribution given by the second term, so that the total passes through a minimum. The effective heat transfer coefficient to the tubes will be slightly lower than that given by Equations 8 and 9, since

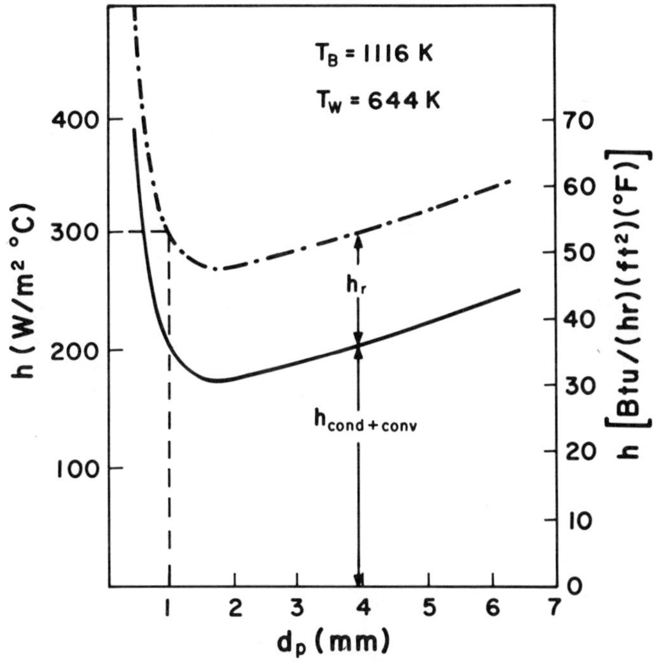

Figure 6. Contributions of the convective and radiative contributions to heat transfer on particle diameter.

the tubes will be shielded periodically from contact with the dense phase by bubbles.

The area of the tubes immersed in the bed is then determined from energy balances on the FBC (Figure 7). An energy input with the coal of 111 MW is needed to provide the desired thermal output of 100 MW and the losses of 9.8 MW with the stack gases (for an exhaust temperature of 450 K) and 1.3 MW with the spent bed solids (assuming no recovery of the sensible heat). Given a recycle of 13 MW to the bed as air preheat, 73 MW must be extracted by coils immersed in the bed in order to maintain the bed temperature at 1116 K. The tube area needed to achieve this is 515 m^2, provided for example by six rows of 50 mm O.D. tubes with a horizontal pitch of 150 mm and a vertical pitch of 125 mm.

Control of bed temperature provides some special challenges. As the load is varied, either the heat transfer coefficient or immersed tube area must be varied correspondingly. This may be achieved over a moderate load change by varying the area of tubes immersed in the bed by varying the height of the bed.

Carbon Combustion Efficiency (Selection of Coal Feed Diameter)

The principal loss of combustible from the bed is by loss of unburned carbon either elutriated from the bed or withdrawn with the bed solids. In order to evaluate the carbon losses it is desirable to consider a balance on the carbon in the bed (34). To simplify the analysis we will first approximate coal as carbon. The governing population balance for the carbon in the bed is:

$$\underbrace{F_o p_o(R)}_{\text{input}} = \underbrace{F_1\left(\frac{W_c}{W_t}\right)p(R)}_{\text{solids withdrawal}} = \underbrace{W_c K(R) p(R)}_{\text{elutriation}} + \underbrace{W_c \frac{d[p(R)\mathcal{R}]}{dR}}_{\substack{\text{net shrinkage}\\ \text{out of size range}}} - \underbrace{\frac{3W_c p(R)\mathcal{R}}{R}}_{\substack{\text{growth within}\\ \text{size range}}} \qquad (10)$$

where F_o = coal feed, kg/s
F_1 = solids withdrawal rate, kg/s
$K(R)$ = elutriation constant for particles of size R, s^{-1}
$p(R)$ = fraction of mass of coal in size range R to R + dR, m^{-1}
$p_o(R)$ = p(R) for feed stream, m^{-1}
R = particle radius, m
$\mathcal{R}$ = rate of change of particle radius, m/s
W_c = carbon loading, kg
W_t = total solids in bed, kg

For the special case of a feed of constant size R_1, i.e. $p_o(R) = \delta(R-R_1)$, for negligible loss of carbon by elutriation and

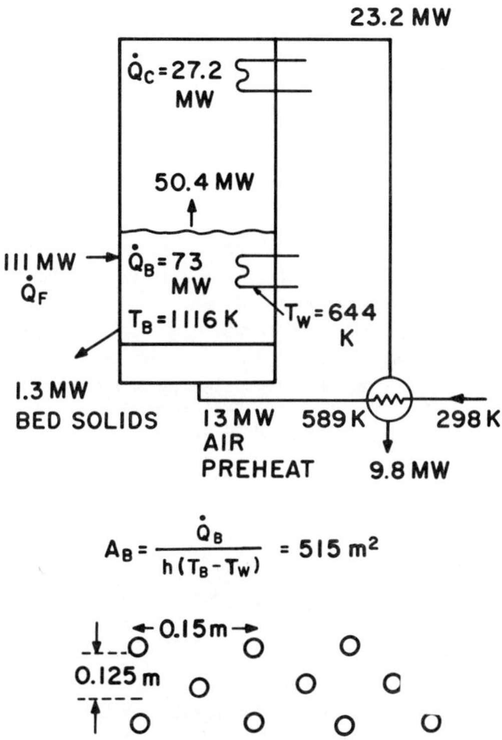

Figure 7. Energy balance on a hypothetical 100 MW_t boiler and the in-bed heat transfer area required to maintain the bed temperature at 1116 K, e.g., 6 rows of 50 mm O.D. tube with a horizontal pitch of 150 mm and a vertical pitch of 125 mm.

with the solids withdrawn from the bed [$K(R) = F_1 = 0$], and for a diffusion limited carbon burning to CO_2 ($\mathcal{R} = -6\mathrm{Sh}\mathcal{D}C_T y_{O_2}/\rho_c R$, where $\mathcal{D}$ is the oxygen diffusivity, C_T the total molar gas concentration, ρ_c the carbon density, and y_{O_2} the oxygen mole fraction), Eq. 10 can be solved for the particle size distribution p(R) and the carbon loading. The solutions are

$$p(R) = \frac{\rho_c}{6\mathrm{Sh}\mathcal{D}C_T y_{O_2}} \frac{F_o}{W_c} \frac{R^4}{R_1^3} \tag{11}$$

and

$$W_c = \frac{\rho_c}{6\mathrm{Sh}\mathcal{D}C_T y_{O_2}} \frac{F_o R_1^2}{5} = \frac{2}{5} F_o t_b \tag{12}$$

where t_b is the burning time of particles of initial radius R_1.

The carbon loading is found to be proportional to coal feed rate and particle burning time. It is a variable of major importance because it determines the carbon loss from the bed and also has an important impact on the emissions of nitrogen oxides. Estimates of the carbon loading can be obtained by evaluating t_b from the relation for a shrinking sphere model of a diffusion-limited oxidation of carbon. It is readily shown that, for particles of density ρ_c and initial diameter d_i

$$t_b = \frac{\rho_c d_i^2}{48\mathrm{Sh}\mathcal{D}C_p} \tag{13}$$

where Sh is the Sherwood Number, $\mathcal{D}$ the oxygen diffusivity, and C_p the oxygen concentration in the emulsion phase.

Although some controversy exists on the correlation of Sherwood Number applicable to fluidized beds, well-defined combustion experiments support the use of the Ranz and Marshall (35) or Frossling (36) correlation with an approximate correction of ε_{mf} to allow for the obstruction to diffusion by the inert particles surrounding the burning char particles (37). Thus

$$\mathrm{Sh} \simeq \varepsilon[2 + 0.6\ \mathrm{Re}^{1/2}\mathrm{Sc}^{1/3}] \tag{14}$$

The oxygen concentration C_p will depend upon the excess air, expressed as a fraction of the stoichiometric air, and the fluid mechanics. Assuming that negligible amounts of oxygen by-pass the bed in bubbles, the oxygen concentration C_p in the bed can be related to the inlet oxygen concentration C_o and excess air by

$$C_p = \frac{C_o}{1 + 1/e} \tag{15}$$

for a well-stirred emulsion gas, and

$$C_p = \frac{C_o}{(1+e)\ln(1+1/e)} \tag{16}$$

for plug flow. With char particles 1mm in diameter, an excess air e of 0.2, a bed temperature of 1116 K, and ε_{mf} of 0.5, the burning times are 185 seconds for a well-stirred gas, and 67 seconds for plug flow, The lower times are more pertinent to AFBC's in which the gas is approximately in plug flow. The carbon loading in the bed is then determined by substituting the coal feed rate (3.83 Kg/s for 111 MW thermal input) and burning time into Eq. 12. A carbon loading of 103 Kg is obtained for the case of plug flow in the gase phase. The fraction of the total bed weight which is in the form of carbon is small. Substituting into

$$\frac{W_c}{W_T} = \frac{W_c}{Ah\rho_s(1-\varepsilon)} \tag{17}$$

the previously determined values of h = 0.7 m, A = 87 m^2 and the density of limestone (ρ_s = 2400 Kg/m^3), we find that the carbon accounts for only 0.14 percent of the total bed weight.

With carbon loading determined, one is in a position to evaluate the carbon loss from the bed. The three principal sources are carbon withdrawn from the bed with spent solids, elutriation of carbon that has burned down to an elutriable size range, and the elutriation of fines produced by attrition.

The fraction $\dot{m}_{c,s}/\dot{m}_F$ of the carbon feed loss with spent solids is determined by the rate $\dot{m}_s$ of spent solid withdrawal, and the fraction (W_c/W_T) of the bed solids in the form of carbon:

$$\frac{\dot{m}_{c,s}}{\dot{m}_F} = \frac{\dot{m}_s}{\dot{m}_F}\frac{W_c}{W_T} \simeq 10^{-3} \tag{18}$$

The fraction $\dot{m}_{c,e}/\dot{m}_F$ lost by elutriation is determined by the elutriation rate constant K(R), the particle size distribution p(R) of carbon in the bed, and the carbon loading W_c (Eq. 10). A first approximation for this loss may be obtained by assuming that that particles of feed diameter d_i shrink to an elutriable size d^*, approximately 300 μm from Figure 4, and are then blown out of the bed. For this simple model

$$\frac{\dot{m}_{c,e}}{\dot{m}_F} = \left(\frac{d^*}{d_i}\right)^3 \simeq (0.3)^2 \simeq 2.7 \times 10^{-2} \tag{19}$$

Another source of elutriated carbon are the fines produced by the attrition of the carbon surface by abrasion. This mechanism for attrition has been studied by Massimilla and coworkers (38-41).

They find that the rate of attrition is proportional to the energy dissipation rate in the bed, which is proportional to $u_o - u_{mf}$, and that it is much higher during oxidation than pyrolysis. They postulate that asperities are produced on the surface of the char during combustion and that these are abraded by collisions with the bed solids. The rate of the carbon surface regression due to attrition is approximately given by

$$\mathcal{R} = -k_a(u_o - u_{mf}) \tag{20}$$

where k_a is about 3.5×10^{-8}. The rate of feed carbon loss due to the elutriation of attrited carbon is then given by

$$\frac{\dot{m}_{c,a}}{\dot{m}_F} = \int_0^{d_i/2} \frac{3p(R)W_c k_a(u_o - u_{mf})}{\dot{m}_F R}\, dR = \frac{3k_a(u_o - u_{mf})}{d_i} t_b \tag{21}$$

For the conditions we have selected, $\dot{m}_{c,a}/\dot{m}_F$ equals 1.1×10^{-2}.

The above calculations show a carbon loss of about 4 percent of the coal feed, primarily as fines produced by carbon attrition or by the shrinkage of the coal feed. As coal particle feed size increases the attritted carbon increases (note t_b in Eq. 21 is proportional to d_i^2) but the elutriated carbon (Eq. 19) decreases. Carbon losses can therefore be minimized by the judicious choice of coal feed size. The simplified model presented above yields the following expression for the optimum size:

$$(d_i)_{opt} = \left[\frac{48 Sh \mathcal{D} Cp (d^*)^3}{\rho_c k_a (u_o - u_{mf})}\right]^{1/4} \tag{22}$$

For the test case, being considered, the optimum coal feed size given by Eq. 22 is 1.6mm. Clearly, the optimum depends upon operating conditions, and also on the model assumptions.

The fines elutriated from the bed are usually captured in a cyclone and recycled to the bed, so that high combustion efficiencies are achievable with AFBC's. Better understanding of the residence time of fines in a bed is needed in order to determine the recycle rate.

The above simplified analysis was intended to provide a feel for the relative importance of the processes that govern carbon loading, and therefore carbon combustion efficiency. More complete treatments of AFBC's are available which consider the detailed population balance equations for the char particles coupled with an oxygen balance (41-50). These treatments have given results which parallel observations on operating AFBC's but are qualified by uncertainties in the flow models and mass transfer models which are used to calculate the transfer of oxygen to the dense phase and by differences in assumptions concerning the combustion of the char.

The refinements that can be made to the model for coal char oxidation are seemingly endless (51). A few of these will be discussed briefly.

Dominant Oxidant. Carbon gasification in fluidized bed combustors can occur by both the CO_2/C and the O_2/C reactions:

$$CO_2 + C = 2CO \tag{23}$$

$$(1 - \frac{x}{2})\ O_2 + C = xCO + (1 - x)CO_2 \tag{24}$$

Avedasian and Davidson (37) in their pioneering work on the modelling of char oxidation in fluidized beds assumed a two-film model for char oxidation, in which CO_2 reacts with carbon at the surface to form CO with the CO being oxidized in a diffusion flame enveloping the particle. Later studies (52) showed that the CO_2/C reaction was too slow at fluidized bed temperatures to contribute significantly to the carbon gasification rate. The primary product of the C/O_2 reaction is believed to be CO, based on molecular beam studies on graphite (53). CO is then converted to CO_2 as it diffuses from the particle surface. A generalized treatment of the carbon oxidation reactions allowing for the two gasification reactions, Eqs. 23 and 24, and the carbon monoxide oxidation in the boundary layer has been presented by Amundson and coworkers (54-58).

External Diffusion vs. Chemical Kinetics. As expected, the rate of carbon oxidation is determined by the combination of external diffusion and the kinetics of the carbon oxidation. Early calculations by Borghi et al (59) determined the contributions of kinetics and external diffusion as a function of bed temperature, particle size, and oxygen concentrations. These show that for particles of 1 mm diameter 1116 K, conditions typical of commercial practice, the diffusion and kinetic resistances are of equal importance. Experimental support for these predictions is provided by Ross and Davidson (60, 61).

Reaction Order. Studies of the reaction of oxygen with carbon at temperatures of interest for AFBC's suggest that it is near zero order in oxygen (62). Most models have been based on an assumed first order reaction but they can be readily modified to accommodate the more realistic lower reaction order (63, 64). The correction for order of reaction will be most important for the prediction of the combustion of recycled fines which are in the size range in which chemical kinetics dominate and for predicting the performance of pressurized fluidized beds.

Pore Structure. The chars produced from coals are highly porous, with internal surface areas of 100 to 1000 m^2/g, and pore radii ranging from nanometer to micron in size. Models of increasing sophistication have been developed to describe the reaction in porous chars (65-71) which allow for diffusion and chemical reaction, pore size distribution, and variation in pore size with increasing extent of reaction. The added refinements are not needed for purposes of calculating char burnout times for most of the carbon in AFBC's, since the external resistance constitutes a significant fraction of the total resistance. Use of such models are, however, desirable when examining details of secondary reactions such as those that influence pollutant formation which occur within the pores, or in evaluating the effect of pore structure on attrition and fragmentation rates.

Particle Temperature Overshoot. The temperature of the burning char particles will run hotter than that of the bed by amounts that depend upon particle size, reactivity, bed temperature. It is determined in part by the heat released at the particle surface due to reaction and in part to the additional heat released by carbon monoxide oxidation near the particle surface (54-58). Measurements for 1.8 to 3.2 millimeter size coke particles burning in a fluidized band of sand at 1173 K increased from the bed temperature at low oxygen concentrations to values 150 to 200 K above the bed temperature for oxygen concentrations approaching that of air (72). Estimation of this temperature rise is important for purposes of evaluating the NO/C reaction and also for prediction of the burnout times of fines.

Carbon Monoxide Oxidation. Analysis of the carbon monoxide oxidation in the boundary layer of a char particle shows the possibility for the existence of multiple steady states (54-58). The importance of these at AFBC conditions is uncertain. From the theory one can also calculate that CO will burn near the surface of a particle for large particles but will react outside the boundary layer for small particles, in qualitative agreement with experimental observations. Quantitative agreement with theory would not be expected, since the theoretical calculations, are based on the use of global kinetics for CO oxidation. Hydroxyl radicals are the principal oxidant for carbon monoxide and it can be shown (73) that their concentration is lowered by radical recombination on surfaces within a fluidized bed. It is therefore expected that the CO oxidation rates in the dense phase of fluidized beds will be suppressed to levels considerably below those in the bubble phase. This expectation is supported by studies of combustion of propane in fluidized beds, where it was observed that ignition and combustion took place primarily in the bubble phase (74). More attention needs to be given to the effect of bed solids on gas phase reactions occuring in fluidized reactors.

Ash-Diffusion Control. A U.S. coal contains on the average about 10 percent by weight of mineral constituents. These yield an ash during combustion which tends to accumulate on the surface of the receding char. Since AFBC's operate below the fusion temperature of ash, the ash either flakes off as the carbon surface recedes or forms a highly porous, loosely-sintered, friable cage about the char. In principle, the ash layer should retard the rate of char combustion (34) but the experimental evidence to date suggests that its effect is small in determining over-all burning rates. It may, however, influence the latter stages of combustion and the achievement of high carbon burnout.

Particle Fragmentation. An added complication in the modelling of coal combustion is provided by the devolatilization step, which determines the yield of the char, its porosity, particle size, and particle number. Massimilla and coworkers (41) have measured the coal fragments produced during pyrolysis. Indirect evidence for fragmentation of various chars during combustion was also provided by Campbell and Davidson (75) who attributed the deviation of the particle size distribution during the combustion of char particles of uniform initial size from the predicted size distribution given by Eq. 11 as being due to fragmentation.

Although a complete model for single particle burning requires detailed information on physical parameters, such as pore structure and chemical parameters, such as intrinsic reactivities of char/oxygen, the near dominance of the external diffusion results in good agreement being obtained between the predictions of burning times obtained using relatively simple models and experimental results. The detailed models will prove useful in resolving issues that cannot be addressed by simpler models, such as the importance of high internal CO concentrations on the augmentation of the reduction of NO by char, on the effect of evolving pore structure on attrition and fragmentation, and on where CO combustion occurs.

Volatile Evolution (Coal Feed Location)

Most models of fluidized-bed combustors have been developed for chars on cokes with negligible volatile content or, if addressing coals, have treated the volatile evolution as being either instantaneous or uniformly distributed throughout the fluidized bed. The coal devolatilization step is, however, of major importance to the operation of AFBC's since the coal must be distributed throughout a bed in a manner that ensures that the volatiles will have access to sufficient air for their complete oxidation.

If the volatile release rate is slow relative to the rate of solid circulation, the assumption of uniform volatile release would be justified. Estimates of the volatile release rate may be obtained from the extensive literature on coal pyrolysis (76).

Using the data of Howard and coworkers (77) Borghi et al (59) have modeled the volatile release rate for AFBC conditions. Their prediction for both the duration of a volatile flame and the subsequent char burnout times have been confirmed by the experimental measurements of Andrei (78). Sample results are shown in Figure 8, which presents the calculated (solid line) and measured (data points) weight loss of 3 millimeter particles burned in 5 percent oxygen stream in a fluidized bed maintained at 1173 K. The initial steep drop in weight is due to volatile evolution, and it is followed by the slower char burnout stage. Thirty percent of the weight of coal is lost in a period of a little under 4 seconds for the 3 mm particle. For 1 mm particles the time for devolatilization is under 1 second. These times can be compared with the estimates of the solid circulation time of 3 to 6 seconds obtained above. The volatile release time is therefore comparable to the solid circulation time and neither the model of instantaneous or distributed volatile release is valid.

Park et al (42) recognized the importance of modeling the volatile release in AFBC's and developed a plume model based on the instantaneous release of volatiles and lateral diffusion using a diffusivity based on tracer dispersion. Their model succeeded in showing the major importance of lateral diffusion within AFBC's in determining feed point location. Stubingdon and Davidson (79) however showed that the combustion of hydrocarbons, which are a valid surrogate for coal volatiles, is governed by molecular diffusion which is considerably lower than the radial diffusion obtained from tracer studies. (The tracer studies apparently provide a measure of the meandering of a laminar stream without mixing at the molecular level needed for reaction). The mechanism for radial dispersion is due to solid and not gas diffusion. A slight variant on the plume model of Park et al (42) is shown in Fig. 9. Coal particles injected at the bottom of the bed are convected upward and diffuse radially as a consequence of bubble motion. Volatiles will be released at a decaying rate for several seconds, depending upon particle size and bed temperature. The radius r of the cross-section over which the volatiles are released is determined by the solid diffusivity $\mathcal{D}_s$. To a first approximation

$$r = 2\sqrt{\mathcal{D}_s t} \tag{25}$$

The correlation proposed by Kunii and Levenspiel (34) will be here used to obtain an estimate for $\mathcal{D}_s$

$$\mathcal{D}_s \simeq \frac{3}{16}\left(\frac{\delta}{1-\delta}\right)\frac{u_{mf}d_b}{\varepsilon_{mf}} \tag{26}$$

The mechanism of solid dispersion is complex and poorly understood so that Eq. 26 represents only a first approximation. For

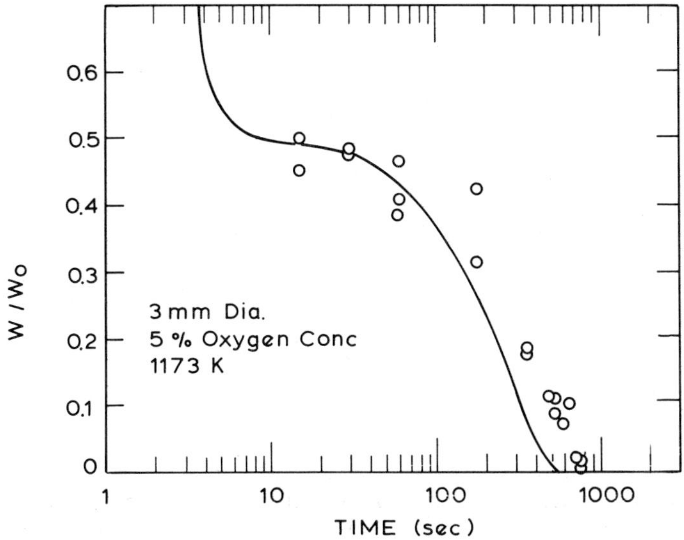

Figure 8. Time-resolved values of the fractional weight retention of coal particles initially 3 mm in diameter burned in a fluidized bed at 1173 K and an oxygen concentration of 5 percent. Solid line calculated from theory, data points experimental.

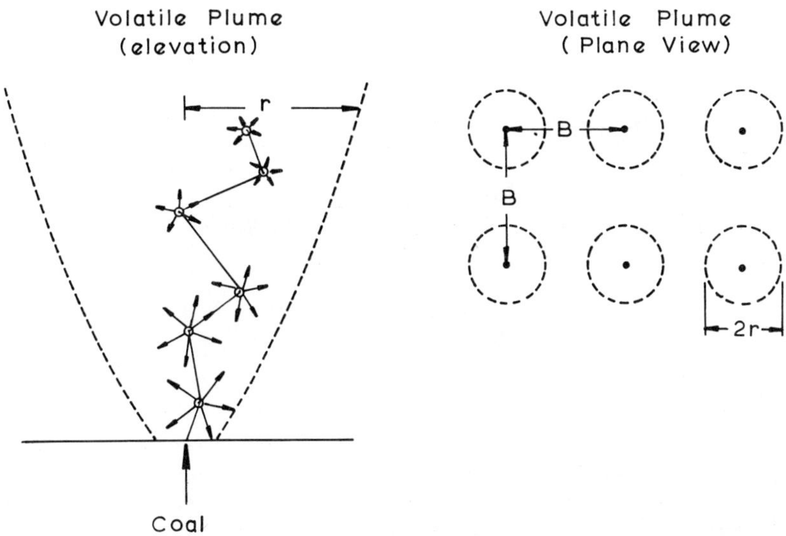

Figure 9. Schematic elevation and plan views of volatile plume formed by coal injected at base of bed.

a void volume δ of 0.3, a mean bubble diameter of 0.5, the value of $\mathcal{D}_s$ = 0.03 m^2/s. The time for the bubble to rise to the surface we have previously shown to be of the order of 1.5 to 3 second. Consequently r $\simeq$ 0.4 to 0.6 m.

The spacing B of coal feed points (Figure 9) should be such that the air flowing in the cross-section πr^2 of the volatile plume should be sufficient to oxidize the volatiles. The fraction f_v of the stoichiometric air requirement needed to consume the volatiles will depend upon the volatile yield and composition (76, 79, 80). A typical value for f_v is about 0.4. A fraction f_a of about 0.1 of the stoichiometric air is used to transport the coal. The oxygen requirements of the volatiles is then satisfied when

$$\frac{\pi r^2}{B^2} = \frac{f_v - f_a}{1+e} \tag{27}$$

For the values selected above B is 1.4 to 2.1 m. This estimate agrees with current commercial practice of spacing feed points at intervals of 1 to 3 m. For our test case, this corresponds to one coal feed point for every 1.8 to 4 MW, or approximately 25 to 55 feed points. The requirement for a large number of feedpoints is one of the shortcomings of AFBC's. Methods of overcoming the problem of multiple feed points include the use of a spreader-stoker to distribute the coal evenly over the surface of the bed or the promotion of lateral mixing by recirculating bed solids.

Emissions of Nitrogen Oxides (Staging of Air Addition)

One of the advantages of AFBC's is that their emissions of nitrogen oxides can be considerably lower than corresponding values from pulverized coal fired boilers. The nitrogen oxides are produced primarily by the oxidation of nitrogen organically bound in the coal (82, 83). Rapid conversion of the fuel bound nitrogen occurs near the distributor place with peak levels of 1000 ppm NO and higher being observed. The NO concentration then decays as a consequence of both heterogeneous and homogeneous reactions.

AFBC's operate in the temperature regime in which ammonia reacts selectively with NO in the presence of oxygen (84):

$$NH_3 + NO + 1/4\ O_2 = N_2 + 3/2\ H_2O \tag{28}$$

Ammonia is evolved with the coal volatiles and may be therefore partially responsible for the reduction of NO that occurs in AFBC's. Because of the inhibition of free radical reactions within the bed this reaction is expected to be most important in bubbles and in the solid disengaging height above the bed. This importance of the NH_3/NO reaction and its partial suppression by bed solids has been demonstrated by the injection of NH_3 into different positions within a fluidized bed combustor (85). These experiments showed that the NO reduction was most effective when the ammonia was injected directly above the bed.

Several heterogeneous reactions contribute to the reduction of NO in AFBC's. Carbon will reduce NO directly:

$$(1 - \frac{x}{2})NO + C = 1/2\ N_2 + xCO + (1-x)CO_2 \tag{29}$$

The rate of the reaction (86-90) is about two orders of magnitude slower than the O_2/C reaction, consistent with the greater strength of the NO bond than that in O_2. The CO/CO_2 ratio in the products of the reaction increases with increasing temperature (86, 87). At low temperatures (850 K), a stable chemisorbed oxygen compled (86) forms and inhibits the reaction. At AFBC temperatures, however, it has been observed that the reaction is accelerated in the presence of oxygen (91). This latter result may be a consequence of the increase in the CO concentration within a char particle as the O_2 concentration is raised. Because the O_2/C reaction is so much faster than the NO/C or the carbon catalyzed CO/NO reaction (86, 91), the situation exists in which the effectiveness factor for the O_2/C reaction is small and little O_2 penetration into char occurs at a time when the effectiveness factor for the NO reduction reactions are near unity. Additional NO reduction reactions that may occur are the CO/NO reaction catalyzed by bed solids (90 - 92) and the reduction of NO by sulfite-containing, partially sulfated limestone (93).

$$CaSO_3 + NO = CaSO_4 + 1/2\ N_2 \tag{30}$$

A complete modeling of NO formation in AFBC's is a formidable task, depending upon the knowledge of the (i) devolatilization chemistry and kinetics of fuel nitrogen which determines the distribution of nitrogen between char nitrogen, ammonia, and other species, (ii) the kinetics of formation of NO by the oxidation of the products of devolatilization, (iii) the kinetics of reduction of NO by NH_3, C, CO, and $CaSO_3$. In order to solve for the NO concentration one needs to also determine the concentration of C, CO, and $CaSO_3$ in the bed and for the temperature overshoot by the carbon. Because of the many chemical reactions involved, and the additional complexity introduced by the fluid mechanics, models of NO formation in beds are only partial. Several have been developed which adequately correlate available experimental data (94-98) but their predictive capability is yet to be proven. The understanding of the dominant mechanism for NO formation and reduction can, however, guide the selection of appropriate control strategy. For example, the NO reduction by char can be accelerated by increasing the carbon loading in the bed by staging the air addition (99-100). But staging the oxygen will increase the carbon loading and correspondingly the formation of carbon fines, so that the reduction in NO may be at the expense of an increase in carbon loss or an increase in recycle of fines.

Freeboard Reactions (Determination of Freeboard Height)

A height of three to four meters above the bed is required to allow solids entrained by the bubble wakes into the freeboard to return to the bed surface. The initial velocity of the solids that splash into the freeboard is 4 to 8 times the bubble rise velocity (100) and the freeboard height is determined by the kinetic energy of the larger particles for which drag forces are relatively unimportant. For a bubble rise velocity of 3.2 m/s (estimated from $u_o - u_{mf} + 0.71\sqrt{gd_b}$ (29) and a bubble diameter of 0.5 m), the maximum height obtained by a particle with an initial velocity six times the bubble rise velocity is 4.3 m, a typical height selected for the freeboard.

The solid concentration in the freeboard is sufficiently high to contribute significantly to the reactions occuring in AFBC's. Reactions of particular importance are those between NO and char, the burnout of carbon monoxide, and the combustion of the recycled fines. Space is not available to cover the subject adequately and the reader is referred to selected literature on the subject (101-104). It is of interest to note, however, that the elutriation rate of larger particles has been found to be greatly augmented by the concentration of fines (103) in a bed, so that the recycle of fines may influence bed behavior in an unobvious manner.

Impact of Immersed Tubes on Flow Regimes

In the above attempt to provide an order-of-magnitude estimate of the factors controlling the design and operation of AFBC's gross approximations were made, particularly in estimating some of the flow and mixing parameters. A number of the correlations used were derived from the literature on bubbling beds of fine particles. It must be emphasized that the flow regimes in beds of coarse particles differ from those in beds of fine particles (105, 106) and that this must be taken into consideration in the selection of the most relevant models.

The early studies of flow in coarse particle systems was for beds without immersed tubes (107, 108). More recent studies have shown that the fluid mechanics and gas and solid mixing can be considerably influenced by the presence of tubes in the bed (109-114). The effect of tubes in the bed is to decrease the pressure fluctuations, presumably due to a reduction in the size of the bubbles. The net consequence of this would be to increase the importance of the slow or cloudless bubble regime (see Figure 4). The data of Fitzgerald, Levenspiel and co-workers (109) show that the impact of tubes is to reduce the dispersion of gas due to the meandering of a tracer gas plume but to increase the diffusion due to turbulent mixing. This augmentation of mixing by the tubes should be taken into account in the modeling of volatile combustion.

Although general treatments of the flow in fluidized beds of coarse particles, in view of the difficulty of the problem, will only evolve slowly, there have been some promising developments. These include the success achieved in correlating heat transfer data (7, 113, 114) and the development of scaling parameters that permit the use of cold flow models to study many of the characteristics of AFBC's (114, 115).

Concluding Comments

The above consideration of the factors that govern the design and operation of the first generation of AFBC's provide an appreciation for the parameters that constrain their performance. There is clearly opportunity for improvement, particularly with respect to reducing the amounts of sorbent used for sulfur capture, reducing the number of coal feed points, reducing the loss of carbon fines, and changing heat load without varying bed temperature. A number of fluidized bed combustors of advanced design have been developed or are under investigation. These include fast fluidized beds, pressurized beds, multi-stage beds, circulating beds, and rotating beds. The fast-fluidized recirculating bed is furthest developed of those advanced concepts, and has already been commercialized.

From a fundamental standpoint, the order of magnitude estimates of many of the parameters in the bed was intended to indicate which processes were important. There is clearly a need for a better understanding of the processes that govern (i) particle circulation and elutriation in large particle systems, (ii) grain growth for purposes of improving stone utilization, (iii) the generation and burnout of fine carbon particles.

An attempt was made in the presentation to show the complex interaction between the different processes occuring in the bed. For example, sorbent utilization may affect NO_x through the formation of $CaSO_3$ and bed pressure drop through the dependence of solid density on extent of sulfation; carbon burning times determine the carbon loading in the bed which in turn governs both the amount of fine char particle generation and the reduction of NO; the carbon fines influence the elutriation rates of coarser particles; the carbon monoxide influences the carbon particle temperature overshoot and also participates in the surface catalyzed reduction of NO. Such interactions add to the challenges of modeling this class of chemical reactors.

Legend of Symbols

A_o bed crossectional area, m^2

B spacing between coal feed points, m

C_o inlet oxygen concentration, $mole/m^3$

C_p average oxygen concentration in emulsion phase, $mole/m^3$

Continued on next page

Legend of Symbols--Continued

C_T total gas-phase molal concentration mole/m^3

d_b bubble diameter, m

d_i initial diameter of char particle, m

d_p bed particle diameter, m

d^* diameter of particles with terminal velocity lower than fluidizing velocity, m

$\mathcal{D}$ gas-phase diffusivity, m^2/s

$\mathcal{D}_s$ diffusion coefficients for solids in emulsion phase, m^2/s

e excess air, fraction of stoichiometric air requirement

f_a fraction of stoichiometric air used as carrier gas for coal feed

f_v oxygen requirement to completely combust the volatiles, fraction of stoichiometric air requirement

F_o coal feed rate, Kg/s

F_1 solids withdrawal rate, Kg/s

g grativational acceleration, m^2/s

h height of bed, m

h convective heat transfer coefficient in Equation 8, W/(m^2)(°C)

h_r radiative heat transfer coefficient, W/(m^2)(°C)

$-\Delta H_c$ heat of combustion, J/Kg

k_a coefficient of attrition defined by Equation 20, dimensionless

K(R) elutriation constant for particles of size R, s^{-1}

K_s rate constant for the SO_2/CaO reaction, cm/s

m fraction of calcium converted to the sulfate

m_∞ value of m approached asymptotically at long times

$\dot{m}_c$ rate of carbon loss from the bed, Kg/s. Subscripts s, e, and a refer to loss with solids withdrawal, by elutriation, and by attrition respectively

Legend of Symbols--Continued

$\dot{m}_F$	coal feed rate, Kg/s
$\Delta\dot{m}_F$	coal feed rate needed to supply power for air fans, Kg/s
$\dot{m}_s$	rate of solids withdrawal from bed, Kg/s
$p(R)$	fraction of mass of coal in size range R to R + dR, m^{-1}
$p_o(R)$	p(R) for coal feed stream, m^{-1}
ΔP	pressure drop across bed, N/m^2
r	radius of volatile plume, m
R	particle radius, m
R_1	particle radius of monodisperse feed, m
Re	Reynolds number based on velocity u_{mf}/ε_{mf} in emulsion phase
$\mathcal{R}$	rate of change of particle radius, m/s
Sc	Schmidt number
Sh	Sherwood number
t	residence time, s
t_b	particle burnout time, s
T	absolute temperature, °K. Subscripts b, w refer to bed and tube wall temperature, respectively
u_{br}	bubble rise velocity, m/s
u_{mf}	minimum fluidizing velocity, m/s
u_o	fluidizing velocity, m/s
u_t	particle terminal velocity, m/s
v_a	volumetric air supply per unit mass of coal, m^3/Kg
W_c	carbon mass in the bed, Kg
W_t	total mass of the bed, Kg
y_{O_2}	mole fraction of oxygen
α	ratio of wake to bubble volume
δ	fraction of bed volume occupied by bubbles

Legend of Symbols--Continued

ε bed void fraction

ε_{mf} bed void fraction at minimum fluidizing velocity

η_e fractional efficiency of conversion of thermal to electrical energy

η_f fan efficiency

ρ_c carbon density, Kg/m^3

ρ_s bed solid density, Kg/m^3

σ Stefan-Boltzmann constant, $W/(m^2)(°K^4)$

τ characteristic heating time of particles, s

Acknowledgments

The above overview has drawn on background developed as part of an interdisciplinary program on fluidized bed combustion at MIT funded by DOE, with partial support from Stone and Webster Engineering Corp. The authors is indebted to his students and colleagues for his education on many aspects of the problem especially to Janos M. Beér, Christos Georgakis, and Leon R. Glicksman whose views are reflected, albeit through a glass darkly, in the paper.

Literature Cited

1. Willis, D.M. Proc. 6th Int. Conf. on Fluidized Bed Combustion 1980, I, 23-29.
2. Zhang, X.Y. Proc. 6th Int. Conf. on Fluidized Bed Combustion 1980, I, 36-40.
3. Gamble, R.L. Proc. 6th Int. Conf. on Fluidized Bed Combustion 1980, II, 307-317.
4. High, M.D. Proc. 6th Int. Conf. on Fluidized Bed Combustion 1980, I, 41-44.
5. Carls, E.L.; Kaden, M.; Smith D.; Wright, S.J.; Jack, A.R. Proc. 6th Int. Conf. on Fluidized Bed Combustion 1980, II, 225-239.
6. Mickley, H.S.; Fairbanks, D.F. AIChE J 1955, 1, 374.
7. Glicksman, L.R.; Decker, N.A. Proc. 6th Int. on Fluidized Bed Combustion 1980, III, pp 1152-1158
8. Hartman, M.; Coughlin, R.W. Ind. Eng. Chem. Process Des. Develop. 1974, 13, 248-253.
9. Fieldes, R.B.; Davidson, J.F. "Reactions of SO_2 with Limestone in a Fluidized Bed; Estimation of Kinetic Data from a Batch Experiment" AIChE 71st Annual Meeting, 1978.
10. Zheng, J.; Yates, J.G.; Rowe, P.N. Chem. Eng. Sci. 1982, 37, 167-174.
11. Pigford, R.L.; Sliger, G. Ind. Eng. Chem. Process Des. Develop. 1973, 12 85-91.
12. Hartman, M.; Coughlin, R.W. AIChE J 1976, 22, 490-498.
13. Borgwardt, R.H. Environ. Sci. Technol. 1970, 4, 59.

14. Ramachandran, P.; Smith J. AIChE J. 1977, 23, 353.
15. Chrostowski, J.W.; Georgakis, C. ACS Symposium Series 1978, 65, 225-237.
16. Fieldes, R.B.; Burdett, N.A.; Davidson, J.F. Trans I Chem E 1979, 57, 276-280.
17. Jonke, A.A.; Vogel, G.J.; Anastasia, L.J.; Jarry, R.L.; Ramaswami, D.; Haas, M.; Schoffstoll, C.B.; Pavlik, J.R.; Vargo, G.N.; Green, R. Argonne National Laboratory Annual Report ANL/ES-CEN-1004 1971, p 25.
18. Roberts, A.G.; Stantan, J.E.; Wilkins, D.M.; Beacham, B.; Hoy, H.R. Inst. Fuel Symp. Series No. 1 1975, 1, D4-1 to D4-11.
19. Hartman, M.; Trnka, O. Chem. Eng. Sci. 1980, 35, 1189-1194.
20. Lee, D.C.; Georgakis, C. AIChE J. 1981, 27, 41.
21. Chen, T.P.; Saxena, S.C. Fuel 1977, 56, 401-412.
22. Bethell, F.V.; Gill, D.W.; Morgan, B.B. Fuel 1973, 52, 12.
23. Glasson, D.R. J. Appl. Chem. 1967, 17, 91-96.
24. Glasson, D.R.; O'Neill, P. 6th Proc. Int. Conf. on Thermal Anal. 1980, 1, 517-22.
25. Shearer, J.A.; Smith, G.W.; Moulton, D.S.; Smyk, E.B.; Myles, W.M.; Swift, W.M.; Johnson, I. Proc. 6th Int. Conf. on Fluidized Bed Combustion 1980, III, 1015-1026.
26. Yang, R.T.; Shing, M-S. AIChE J. 1979, 25, 811-819.
27. Battcock, W.V.; Pillai, K.K. Proc. 5th International Conference on Fluidized Bed Combustion 1977, II, 642-649.
28. Merrick, D.; Highley, J. AIChE Symp. Series No. 137 1973, 70, 366-378.
29. Davidson, J.E.; Harrison, D. "Fluidized Particles" Cambridge University Press, England, 1963.
30. Lord, W.K., Sc.D. Thesis in Mechanical Engineering, M.I.T., 1983.
31. Grace, J.R.; Clift, R. Chem. Eng. Sci. 1973, 29, 327.
32. La Nauze, R.D., Fluidized Combustion: Systems and Applications 2nd Int. Conf. Fluid. Combust., J. Inst. Fuel, London, 1980.
33. Darton, R.C.; La Nauze, R.D.; Davidson, J.F.; Harrison, D. Trans. I Chem. E 1977, 55 274-279.
34. Kunii, D.; Levenspiel, O. "Fluidization Engineering", John Wiley and Sons, New York, 1969.
35. Ranz, W.E.; Marshall, W.R. Chem. Eng. Prog. 1952, 48, 141-146, 173-180.
36. Frössling, N. Gerlands Beitr. Geophys. 1938, 52, 170-216.
37. Avedasian, M.M.; Davidson, J.P. Trans. Inst. Chem. Eng. 1973, 51, 121-131.
38. Donsi, G.; Massimilla, L.; Russo, G.; Stecconi, P. Seventeenth Symposium (International) on Combustion 1979, p.205.
39. Donsi, G.; Massimilla L.; Miccio, M.; Russo, G., and Stecconi, P. Comb. Sci. Technol. 1979, 21, pp 25-33.
40. Donsi, G.; Massimilla, L.; Miccio, M. Combust. Flame 1981, 41, 57-69.
41. Chirone, R.; Cammarota, A.; D'Amore, M.; Massimila, L.

Nineteenth Symposium (International) on Combustion 1983, pp 1213-1220.
42. Park, D.; Levenspiel, P.; Fitzgerald, T.J. Fuel 1981, 60, 295-306.
43. Yagi, S.; Kunii, D. Fifth Symposium (International) on Combustion, pp 231-244.
44. Saxena, S.C.; Rehmat, A. Proc. 6th Int. Conf. on Fluidized Bed Combustion 1980, 1138-1149.
45. Horio, M.; Wen, C.Y. AIChE Symp. Ser. No. 161 1978, 73, 9-21.
46. Beér, J.M.; Massimilla, L.; Sarofim, A.F. Inst. Fuel Symp. Ser. No. 4 1980, IV (5), 1-10.
47. Wells, J.W.; Krishnan, R.P.; Ball, S.J. Proc. 6th Int. Conf. Fluidized Bed Combustion 1980, 3, 773-783.
48. Congalidis, J.P.; Georgakis, C. Chem. Eng. Sci. 1981, 36, 1529-1546.
49. Gordon, A.L.; Amundson, N.R. Chem. Eng. Sci. 1976, 31, 1163-1178.
50. Bukur, D.B.; Amundson, N.R. Chem. Eng. Sci. 1981, 36, 1239-1256.
51. Essenhigh, R.H., "Chemistry of Coal Utilization"; 2nd Suppl. Vol.; Elliott, M.A.; Ed.; Wiley-Interscience: New York, 1981, Chapter 19.
52. Patel, M.S. Ph.D. Thesis, University of Cambridge, Cambridge, England, 1979.
53. Olander, D.R.; Sickhaus, W.; Jones, R.; Schwarz, J.A. J. Chem. Phys. 1972, 57, 408.
54. Caram, H.S.; and N.R. Amundson, Ind. Eng. Chem. Fundam. 1977, 16, 171-181.
55. Mon, E.; Amundson, N.R. Ind. Eng. Chem. Fundam. 1978, 17, 313-321.
56. Mon, E.; Amundson, N.R. Ind. Eng. Chem. Fundam. 1980, 19, 243-250.
57. Sundaresan, S.; Amundson, N.R. Ind. Eng. Chem. Fundam. 1980, 19, 344-351.
58. Sundaresan, S.; Amundson, N.R. Ind. Eng. Chem. Fundam. 1980, 19, 351-357.
59. Borghi, G.G.; Sarofim, A.F.; Beér, J.M. AIChE 10th Annual Meeting, New York, 1976.
60. Ross, I.B.; Davidson, J.F. Trans I Chem E 1979, 57, 215.
61. Ross, I.B.; Davidson, J.F. Trans. Inst. Chem. Eng. 1982, 60, 108.
62. Froberg, R.W.; Essenhigh, R. Seventeenth Symposium (International) on Combustion 1979, 179-187.
63. Stanmore, B.R.; Jung, K. Trans. Inst. Chem. Eng., 1980, 58, 66-68.
64. Calleja, G.; Sarofim, A.F.; Georgakis, C. Chem. Eng. Sci. 1981, 36, 919-929.
65. Smith, I.W., Combust. Flame 1971, 17, 313-314.
66. Smith, I.W.; Tyler, R.J. Fuel 1972, 51, 312-321.

67. Simons, G.A. Combust. Sci. Technol. 1979, 19, 227-235.
68. Simons, G.R. Nineteenth Symposium (International) on Combustion, 1983, pp 1067-1084.
69. Gavalas, G.R. Combust. Sci. Technol. 1981, 24 197-210.
70. Gavalas, G.R. AIChE J 1980, 26, 577-586.
71. Zygourakis, K.; Arri, L.; Amundson, R. Ind. Eng. Chem. Fundam. 1982, 21, 1-12.
72. Ross, I.B.; Patel, M.S.; Davidson, J.F. Trans. Inst. Chem. Eng. 1981, 59, 83-88.
73. Chaung, T.Z.; Walsh, P.M.; Sarofim, A.F.; Beér, J.M. 2nd Specialists Meeting on Oxidation, Combustion Institute, 1982
74. Dennis, J.S.; Hayhurst, A.N.; Mackey, I.G. Nineteenth Symposium (International) on Combustion, 1983, pp 1205-1212.
75. Campbell, E.K.; Davidson, J.F. Inst. Fuel Symp. Ser. 1975, 1, 1, A2.1-A2.9.
76. Howard, J.B. "Chemistry of Coal Utilization"; 2nd Suppl. Vol. Elliott, M.A.; Ed; Wiley-Interscience: New York, 1981, Chapter 12.
77. Anthony, D.B.; Howard, J.B.; Hottel, H.C.; Meissner, H.P. Fifteenth Symposium (International) on Combustion 1975, pp 1303-1316.
78. Andrei, M., S.M. Thesis, MIT, Cambridge, MA, 1979.
79. Stubington, J.F.; Davidson, J.F. AIChE 1981, 27, 59-65.
80. Suuberg, E.M.; Peters, W.A.; Howard, J.B. Seventeenth Symposium (International) on Combustion, 1979, pp 117-128.
81. Solomon, P.R.; Colket, M.B. Ibid pp 131-140.
82. Jonke, A.A.; Carls, E.L.; Jarry, R.L.; Haas, M.; Murphy, W.A.; Schoffstoll, C.B. Argonne National Laboratory Annual Report ANL/ES-CEN-1001, 1969, pp 32-36.
83. Pereira, J.F.; Beér, J.M.; Gibbs, B.M.; Hedley, A.B. Fifteenth Symposium (International) on Combustion 1975, pp 1149-1155.
84. Lyon, R.K., Int. J. Chem. Kin. 1976, 8, 315-318.
85. Hampartsoumian, E.; Gibbs, B.M. Nineteenth Symposium (International) on Combustion 1983, pp 1253-1262.
86. Chan, L.K.; Ph.D. Thesis, MIT, Cambridge, MA , 1977.
87. Levy, J.M.; Chan, L.K.; Sarofim, A.F.; Beér, J.M. Eighteen Symposium (International) on Combustion 1981, pp 111-120.
88. Furasawa, T.; Kunii, D.; Yamada, N.; Oguma, A. Int. Chem. Eng. 1980, 20, 293-244.
89. Kunii, D.; Wu, K.T.; Furasawa, T. Chem. Eng. Sci. 1980, 35, 170-177.
90. de Soete, G.G.; Revue Générale de Thermique, 1980, 19, 545.
91. Furusawa, T.; Tsunoda, M.; Kunii, D. ACS Symposium Series 1982, 196, 347-357.
92. de Soete, G.G. Sixth Members Conference, International Flame Research Foundation 1980, 2, 299.
93. Skopp, A.; Hammons, G.A., "NO_x Formation and Control in Fluidized-Bed Coal Combustion Processes" ASME Winter Ann. Meeting, 1971.

94. Horio, M.; Mori S.; Muchi, I. Proc. 5th FBC Conference 1977, II, pp 605-623.
95. Pereira, F.J.; Beér, J.M., 2nd European Symposium on Combustion 1975, p 339.
96. Beer, J.M.; Sarofim, A.F.; Lee, Y.Y. J. Inst. Energy 1981, 54, 38-47.
97. Wells, J.W.; Krishnan, R.P.; Ball, S.J. 6th Int. Conf. on Fluidized Bed Combustion 1980, III, pp 773-783.
98. Walsh, P.M.; Chaung, T.Z.; Dutta, A.; Beér, J.M.; Sarofim, A.F. Nineteenth Symposium (International) on Combustion, 1983, pp 1281-1289.
99. Gibbs, B.M.; Pereira, F.J.; Beér, J.M. Sixteenth Symposium (International) on Combustion 1977, pp 461-472.
100. George, S.E.; Grace, J.R. AIChE Symposium Series 1978, 176 (74), 67-74.
101. Wen, C.Y.; Chen, L.H. AIChE J. 1982, 28, 117-128.
102. Yates, J.G.; Rowe, P.N. Trans. Inst. Chem. Engrs. 1977, 55, 137-142.
103. Geldart, D.; Cullinan, J.; Georghiades, S.; Gilvray, D.; Pope, D.J. Trans. I Chem. E 1979, 57, 269-275.
104. Martens, F.J.A.; Op den Brouw, H.; Van Koppen, C.W.J. 7th Int. Conf. on Fluidized Bed Combustion 1982, II, 1054.
105. Catipovic, N.M.; Jovanovic, G.N.; Fitzgerald, T.J. AIChE J 1978, 24, 543-546.
106. Bar-Cohen, A.; Glicksman, L.; Hughes, R. Proc 5th Int. Conf. on Fluidized Bed Combustion 1977, III, 458-471
107. Cranfield, R.R.; Geldart, D. Chem. Eng. Sci. 1974, 29, 935-947.
108. Mc Grath, L.; Streatfield, R.E. Trans. Inst. Chem. Eng. 1971, 49, 70-79.
109. Jovanovic, G.N.; Catipovic, N.M.; Fitzgerald, T.J.; Levenspiel, O. "Fluidization", Grace, J.R.; Matsen, J.M., Eds.; Plenum, 1980, pp 325-332.
110. Cranfield, R.R., AIChE Symposium Series Vol. No. 176 1978, 74, 54
111. Fitzgerald, T.J.; Catipovic, N.; and Jovanovic, G. Proceeding of Fifth International Conference on Fluidized Bed Combustion 1977, III, 135-152.
112. Canada, G.S.; McLaughlin, M.H.; Staub, F.W. AIChE Symposium Series No. 176, 1978, 74, 27-37.
113. Catipovic, N.M.; Jovanovic, G.N.; Fitzgerald, T.J.; Levenspiel, O. "Fluidization" Grace, J.R.; Matsen, J.M., Eds., Plenum, 1980, pp 225-234.
114. Staub, F.W.; Kuwata, M.; Ku, A.C,; and Wood, R.T. Proc. 6th Int. Conf. on Fluidized Bed Combustion 1980, III, 784-790.
115. Fitzgerald, T.J.; Crane, S.D. Ibid, pp 815-820b.

RECEIVED May 2, 1983

5

Current Problems in Polymerization Reaction Engineering

W. HARMON RAY

University of Wisconsin, Department of Chemical Engineering, Madison, WI 53706

Polymerization reaction engineering is an important area in the process industry with many diverse and challenging design problems. In this survey, an introduction to some of the key design difficulties is followed by several specific examples illustrating some of the intriguing and exotic phenomena arising routinely in polymerization reactors. A survey of the recent literature indicates a strong upsurge of interest in these problems.

Currently some 200 billion lbs (~100 million metric tons) per year of synthetic polymers are produced in the world in a wide variety of polymerization reactors. The world capacity and expected rate of growth of the 10 largest volume polymers is shown in Table I. The ultimate use of these products ranges from synthetic clothing to latex paints. The total world sales of the raw polymer is now approaching \$100 billion/year and the sales of the final product after processing, molding, compounding, etc. is many times this figure. Thus the field of polymerization reaction engineering is of significant economic importance.

Up until a decade or two ago, polymers were largely speciality materials, manufactured in batch reactors from faithfully followed recipes scaled up from the chemists beaker. However, with the growth of demand and increased price competition (particularly for high volume commodity polymers), more efficient polymerization methods have been required. Manufacturers of high volume polymers are now moving to fewer product lines, more uniform product and the use of continuous reactors. Similarly, producers of relatively low volume, high quality (and value) polymers are finding that their competitive edge comes from a deeper understanding of the relationship between polymerization conditions and product quality.

Although this change from traditional polymerization methods requires a much better understanding of the physical and chemical

0097–6156/83/0226–0101\$09.25/0

Table I. World Capacity and Expected Growth of Largest Volume Plastics (1)

Rank			Capacity, millions of lb per yr.		
1980s	1981	Resin	Mid-1980s	1981	Change
1	2	Polyethylene, low-density	38,047	31,702	20.0%
2	1	Polyvinyl chloride	35,339	32,430	8.9
3	3	Polyethylene, high-density	22,895	18,071	26.7
4	5	Polypropylene	17,809	15,631	13.9
5	4	Polystyrene	17,626	16,491	6.9
6	10	Polyethylene, linear low-density	8,081	2,669	202.8
7	6	Acrylonitrile-butadiene-styrene	4,662	4,374	6.6
8	8	Isocyanates[a]	4,270	3,634	17.4
9	7	Polyols[a]	4,255	4,246	0.2
10	9	Unsaturated polyesters	3,538	3,516	0.6
		Other[b]	8,006	7,359	8.8
		TOTAL	164,528	140,123	17.4%

a Used as indicators for derivative polyurethanes. b includes acrylics, amino resins, cellulosics, fluoropolymers, phenolics, polyacetals, polycarbonates, polyphenylene oxide, styrene-acrylonitrile.

phenomena in the polymerizing medium, (and this requires the services of more well-trained polymerization reactor engineers), the economic benefits from such improved engineering can be enormous. As only one example, new processes for linear low-density polyethylene (such as the continuous fluidized bed process of Union Carbide (2)) present attractive alternatives to the traditional high pressure processes. These new processes can operate at 300-1000 psig rather than the traditional 30,000-50,000 psig and thus can reduce the plant capital expenditure by 50% and energy consumption by 25%. This type of new low density polyethylene process is making significant progress against the older, more well established technology. Because polymerization reaction engineering is a relatively new field, it seems probable that many other traditional polymerization processes can be comparably improved through the application of polymerization reaction engineering.

In preparing this review of polymerization reaction engineering the author is fortunate that there have been so many books and survey articles to appear in the literature recently (cf. Table II). These, taken together, provide good detailed coverage of the essentials of the field; therefore, extensive delegation of detail will be made to these surveys. Foremost

Table II. Some Recent Books and Survey Articles on Polymerization Reaction Engineering

Author(s)	Topic	Reference
Keane (1972)	Single Phase Polymerization	3
Ray (1972)	Mathematical Modelling	4
Keii (1972)	Ziegler-Natta Polymerization	5
Albright (1974)	Polymerization Processes (Monograph)	6
Min and Ray (1974)	Emulsion Polymerization	7
Gerrens (1976)	Polymerization Reactions and Reactors	8
Bouton and Chappelear (1976)	Continuous Reactors (ed. volume)	9
Piirma and Gardon (1976)	Emulsion Polymerization (ed. volume)	10
Ray and Laurence (1977)	Polymerization Reaction Engineering	11
Poehlein and Dougherty (1977)	Continuous Emulsion Polymerization	12
Schildknecht and Skeist (1977)	Polymerization Processes (ed. values)	13
Henderson and Bouton (1979)	Polymerization Reactors (ed. volume)	14
Boor (1979)	Olefin Polymerization	15
Gerrens (1980, 1981, 1982)	Polymerization Technology	16-18
Odian (1981)	Polymerization Kinetics	19
Ray (1981)	Polymerization Reactor Dynamics	20
Bassett and Hamielec (1982)	Emulsion Polymerization (ed. volume)	21
Piirma (1982)	Emulsion Polymerization (ed. volume)	22
El-Aasser and Vanderhoff (1982)	Emulsion Polymerization (ed. volume)	23
Sebastian and Biesenberger (1983)	Polymerization Engineering	24
Quirk (1983)	Ziegler-Natta Polymerization	25

among these are the recent reviews of Gerrens beginning with his outstanding plenary lecture of the 1976 ISCRE meeting (8). This has been updated and expanded in the recent chapter on polymerization technology (16) which has nearly 500 references. Just recently Gerrens has summarized this review in a superb discussion of polymerization reactors (17,18). The novice just beginning work in this area, as well as the experienced polymerization reaction engineer, will both find Gerrens' surveys profitable reading.

In this paper, we shall first provide a broad brush perspective of the field and then focus on some interesting unsolved problems which are the subject of current research.

Polymerization Kinetics and Reactor Design

Much has been written about polymerization kinetics and the essential steps are shown in Table III for the three principal types of mechanisms (free radical, ionic and coordination, and

Table III. Major Classes of Polymerization Reactions

I. Monomer Addition:

A. Free Radical

$$\left.\begin{array}{l} I \xrightarrow[kd]{} 2R\cdot \\ R + M \xrightarrow[k_i]{} P_1 \end{array}\right\} \text{initiation}$$

$$P_n + M \xrightarrow[k_p]{} P_{n+1} \qquad \text{propagation}$$

$$P_n + A \xrightarrow[k_f]{} P_1 + M_n \qquad \text{chain transfer}$$

$$\left.\begin{array}{l} P_n + P_m \xrightarrow{k_{td}} M_n + M_n \quad \text{disproportionate} \\ P_n + P_m \xrightarrow{k_{tc}} M_{n+m} \quad \text{combination} \end{array}\right\} \text{termination}$$

(sometimes chain branching reactions also occur)

B. Ionic (anionic, cationic, coordination)

$$I + M \xrightarrow[k_i]{} P_1 \qquad \text{initiation}$$

$$P_n + M \xrightarrow[k_p]{} P_{n+1} \qquad \text{propagation}$$

(sometimes chain transfer or termination mechanisms also occur)

II. Polymer Coupling (polycondensation, crosslinking)

$$P_n + P_m \xrightarrow[k_p]{} P_{n+m} \qquad \text{propagation}$$

condensation polymerization). However, there is not comprehensive understanding of the kinetics of even these simple polymerizations, particularly with heterogeneous systems. The reader should consult the surveys of Table II for a discussion of the remaining interesting problems in kinetics.

As important as kinetic mechanism are the phase changes that occur in polymerization. Only a small fraction of polymerizations are carried out only in one phase; thus thermodynamics, heat and mass transfer, and the kinetics of the phase change itself all play a role in determining the properties of the product polymer. Table IV indicates the principal types of kinetic mechanisms and reaction media which arise in polymerization reactors. Each of these classes of systems has its own peculiar problems so that polymerization reactor design can often be much more challenging than the design of reactors for short chain molecules.

To illustrate some of the challenging problems of polymeriza-

TABLE IV

KINETIC MECHANISMS AND REACTION MEDIA EMPLOYED IN POLYMERIZATION REACTORS

Reaction medium	Kinetic mechanism			
	Free radical	Ionic	Condensation	Coordination catalysis
1. Homogeneous (bulk and solution)	Vinyl polymers (Styrene, LDPE)	Polyethers (Ethylene oxide)	Polyesters, Polyamides (PET, Nylon)	Polyolefins with soluble catalyst (ethylene-propylene copolymers)
2. Heterogeneous				
a. Emulsion	Vinyl polymers (Styrene, PVC)	-	-	-
b. Suspension	Vinyl polymers (Styrene, PVC)	-	-	-
c. Precipitation (bulk and solution)	Vinyl polymers (PVC)	Polyacetals (formaldehyde) Vinyls (iso-butylene-butyl rubber)	Polyamides (Nylon interfacial poly.)	Polyolefins with insoluble catalyst (Liquid and gas phase processes for HDPE, LLDPE, polypropylene)

tion reactor design, let us list some of the issues which must be faced:

1. Kinetics and Phase Behavior - Table IV represents a simplified picture of the situation; however, some polymerizations go through several phase changes in the course of the reaction. For example, in the bulk polymerization of PVC, the reaction medium begins as a low viscosity liquid, progresses to a slurry (the PVC polymer, which is insoluble in the monomer, precipitates), becomes a paste as the monomer disappears and finishes as a solid powder. As might be expected, modelling the kinetics of the reaction in such a situation is not a simple exercise.
2. Material Mixing and Conveying - With such a variety of morphologies, there can be serious non-idealities in micromixing and macromixing, particularly at high conversions. Often a large amount of mechanical energy is required for mixing and conveying. Sometimes, sticking and fouling of the reactor surfaces by polymer is a difficult design problem.
3. Heat Removal - Most polymerizations release large amounts of heat as monomer is converted to polymer (cf. Table V). In addition, the mechanical energy required for mixing may be converted to heat under highly viscous conditions. Removal of this heat is often difficult for high conversion polymerization because of high viscosity, heat transfer surface fouling, and change of phase during reaction. In many industrial situations, disastrous reactor runaway is an ever present potential hazard because of these heat removal difficulties. This presents a great challenge to the process control engineer as well as to the reactor designer.

TABLE V

HEATS OF POLYMERIZATION FOR SOME COMMON MONOMERS

	cal/mole at 25°C
Ethylene	-21.2
Propylene	-19.5
Butadiene	-17.6
Styrene	-16.7
Vinyl chloride	-22.9
Vinylidene chloride	-18.0
Vinyl acetate	-21.2
Methyl acrylate	-18.5
Methyl methacrylate	-13.2
Acrylonitrile	-18.4
Formaldehyde	- 7.4

4. Quality of Polymer Product - Product quality is a much more complex issue in polymerization than in more conventional short chain reactions. Because the molecular architecture of the polymer is so sensitive to reactor operating conditions, upsets in feed conditions, mixing, reactor temperature, etc. can alter critical molecular properties such as molecular weight distribution, polymer composition distribution, chain sequence distribution, degree of chain branching, and stereoregularity. In addition, the morphological form of the polymer is often a key quality variable. For example, the particle size distribution in emulsion, suspension, and precipitation polymerization can be a crucial product specification. One of the greatest difficulties in achieving quality control of the polymer product is that the actual customer specifications may be in terms of non-molecular parameters such as tensile strength, crack resistance, temperature stability, color or clarity, absorption capacity for plasticizer, etc. The quantitative relationship between these product quality parameters and reactor operating conditions may be the least understood area of polymerization reaction engineering. Table VI summarizes both types of quality control measures. Because of the lack of on-line measurements for most of these product quality variables (molecular or otherwise), control of polymerization reactors is a special challenge.

In subsequent sections, examples will be used to illustrate some of the design considerations listed here.

Although a number of the surveys listed in Table II (e.g.; 6,11,13,16-18) provide a treatment of polymerization reactors and processes in industrial use, Gerrens has provided an outstanding summary of this information in his recent survey (18). Table VII, taken from Gerrens' review provides an indication of the variety of reactor configurations found in industrial practice.

Because the state of the literature through 1980 is so well covered by the surveys listed in Table II, only very recent work on polymerization reaction engineering will be provided here as a supplement. This is categorized by topic in Table VIII. This large amount of literature (which is not exhaustive) over the last 18 months is indicative of the recent explosion of interest in these problems.

Table VI. Some Measures of Polymer Product Quality

<u>END USE PROPERTIES</u>

- Flow Properties (Film Blowing, Molding, etc.)
- Strength
- Stress Crack Resistance
- Color, Clarity
- Melting Point
- Corrosion Resistance
- Abrasion Resistance
- Density
- Impact Resistance
- Temperature Stability
- Swellability
- Plasticizer Uptake
- Spray Drying Characteristics
- Coating and Adhesion Properties

<u>MOLECULAR ARCHITECTURE</u>

Average Molecular Weight and Molecular Weight Distribution (or Melt Index, Viscosity, etc.)

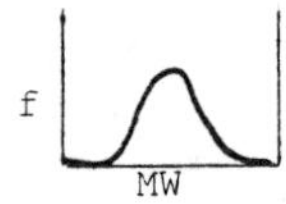

Polymer Composition and Composition Distribution

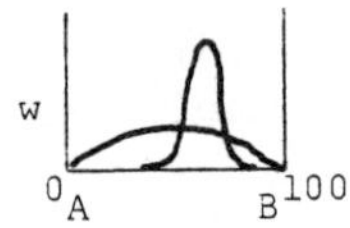

Chain Sequence Distribution

-A-A-B-A-A-A-B-B-

Degree of Chain Branching

a. Linear:

```
 | | | | | |
-C-C-C-C-C-C-
 | | | | | |
```

b. Branched:

```
-C-C-C-C-C-C-
   |     |
   C     C
   |     |
```

Stereoregularity (Tacticity)

a. Isotactic:

```
-C-C-C-C-C-C-C-
 |   |   |   |
 A   A   A   A
```

b. Syndiotactic:

```
     A
     |
-C-C-C-C-C-
 |       |
 A       A
```

c. Atactic:

```
     A
     |
-C-C-C-C-C-C-C-
 |       |   |
 A       A   A
```

Average Particle Size Distribution

Partial Porosity and Surface Area

Table VII. Reactor Configurations Found in Industrial Practice.

	Polymerization reactions	Monomer linkage: with termination — Solution polymerization	with termination — Precipitation polymerization	with termination — Bead polymerization	with termination — Emulsion polymerization	without termination — Solution polymerization	without termination — Precipitation polymerization	Polymer linkage — Solution or melt-polycondensation	Polymer linkage — Interfacial polycondensation	Polymer linkage — Solid-phase polycondensation
BR	Batch stirred tank	1	3	4	5	7	9	10	11	
	Semibatch	2			6	8				
	Piston	12								
	Tumbler									13
	Batch stirred tank + filter press	14								
	Batch stirred tank + autoclave with gate paddle nuscer		15					17		
	2-Component mixer + mold					16		20		
CPFR	Continuous tube	18								
	Continuous tube with distributed bead	19								
	Mixer + conveyor						21	22		
	Extruder						23	24		
	Tower or tower cascade	25						26		27
Cascade	Stirred tank + tower	28						29		
	Stirred tank cascade				30	31	32	33		
	Rotating ring disc							34		
CSTR	(Double) Loop						35			
	Continuous stirred tank	36	37		38		39			
	Fluidized bed						40			

Table VIII. Some Very Recent Work in Polymerization Reaction Engineering

Topic	Principal Research Groups	References
Gel or Trommsdorf Effect in Free Radical Polymerization	Hamielec, Ito, O'Driscoll, Sundberg, Tirrell	(26-48)
Copolymerization	Biesenberger, Johnson, Tirrell	(49-60)
Polycondensation	Gajewski, Gupta, Kaneko, Kumar, Mashelkar, Rafler	(61-94)
Olefin Polymerization with Coordination Catalysts	Bohm, Brockmeier, Hogan, Keii, Pino, Ray, Reichert, Zakharov	(95-124)
Kinetics of High Pressure (Low Density) Ethylene Polymerization	Ehrlich, Jovanovich, Luft, Yamamoto	(125-135)
CSTR for Polymerization	Chen, Fan, Meira, Ray	(136-147)
Tubular Polymerization Reactors	Janssen, Joosten, Middleman, Nigam, Ottino, Ray	(148-166, 246)
Reaction Injection Molding (RIM)	Gogos, Kamal, Macosko, Ottino, Throne, Tirrell	(167-184)
Process Dynamics, Optimization, and Control	Chen, Johnson, Laurence, MacGregor, Ray, Tirrell	(20, 50-54, 136-139, 142-143, 145-147, 185-205)
Suspension Polymerization	Reichert	(206-217)
Emulsion Polymerization	Bassett, Brooks, Guillot, Hamielec, Kiparissides, Nomura, Piirma, Ray, Sundberg, Ugelstad, Vanderhoff	(21-23, 160, 201-205, 218-252)

Some Current Research Problems

To provide an illustration of some of the challenges of polymerization reaction engineering, we shall discuss here a few intriguing but practically important research problems which arise in the design of polymerization reactors. These examples reflect the author's own interest and are selected from research projects currently underway at the University of Wisconsin.

Solution Polymerization in a CSTR. Although many polymerization reactors in use by industry have the residence time distribution of a CSTR, they may not, at first glance, have the appearance of a CSTR (cf. Figure 1). Nevertheless, CSTR models, perhaps with some allowance for imperfect micromixing, are successfully employed to describe these reactors. Thus the behavior of the CSTR is of great practical interest.

Recent studies of CSTR's for bulk or solution free radical polymerization indicate the possibility of multiple steady states (e.g.; 20,137-139,145-147). The fact that this occurs should not be surprising given the large heat evolution and difficult heat transfer characteristic of polymerization reactors. However, there are dramatic new twists to these phenomena even in simple solution polymerization. For example, autocatalytic kinetics lead to the existence of multiple steady states even under isothermal reactor conditions. Figure 2 (taken from (145)) illustrates this for the case of methyl methacrylate polymerization in ethyl acetate solvent carried out at the boiling point of the reacting liquid. Note that the model predictions and the steady state experiments agree that there is ignition to a high conversion, inoperable, steady state at a certain residence time for solvent fractions less than ~40%. The upper steady state is inoperable due to the extremely high viscosities at such high conversions. In the laboratory, ignition which was allowed to proceed to this upper steady state resulted in a reactor filled with solid polymer. The practical consequence of this for the industrial reactor is that even with perfect temperature control (an unlikely event after ignition has occurred), the reactor will ignite into a "runaway" condition for certain combinations of solvent concentration and reactor residence time. If only temperature measurements are being taken, the operator would not be aware of the runaway until the agitator motor blew a fuse, or the greatly increased rate of reaction exceeded the cooling system capacity and caused a sharp increase in temperature and pressure in the reactor.

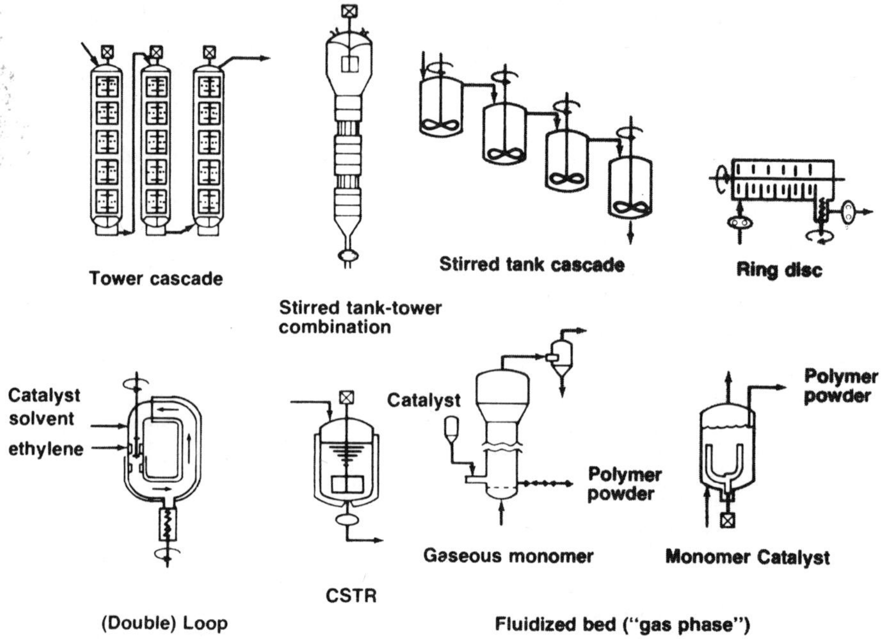

Figure 1. Continuous stirred tank reactor types in use by industry (18).

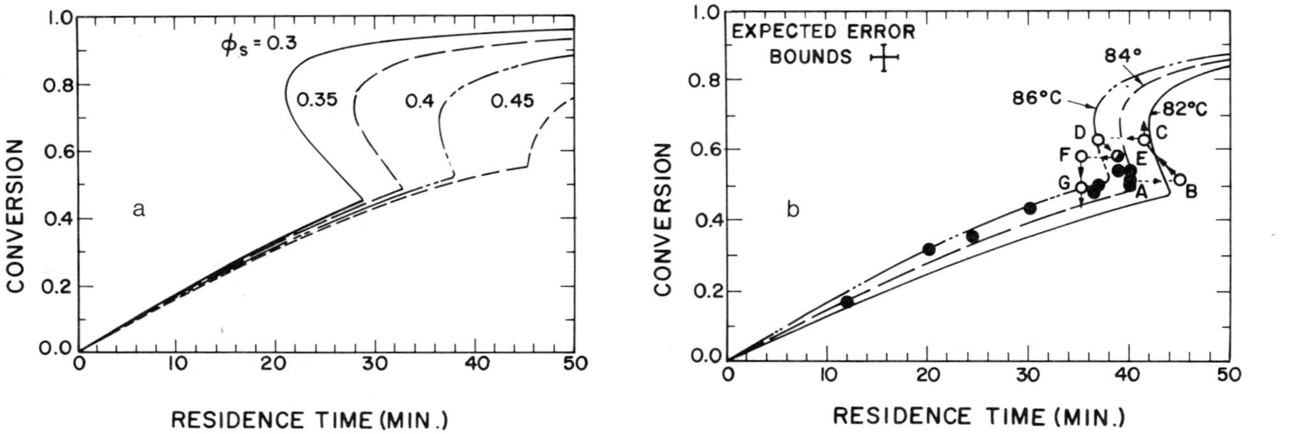

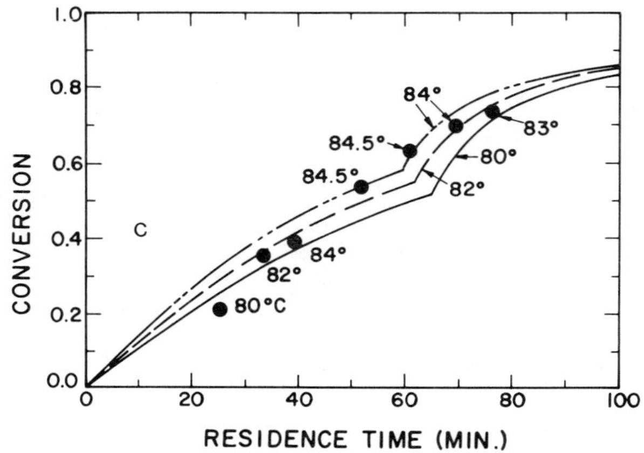

Figure 2. Isothermal polymerization of methyl methacrylate in a CSTR (145). a. Predicted steady-state monomer conversion vs. reactor residence time for the solution polymerization of MMA in ethyl acetate at 86 °C. b. Steady-state and dynamic experiments for the isothermal solution polymerization of MMA in ethyl acetate (solvent fraction 0.4) (●) steady states, (o) non-steady states. c. Steady-state conversion vs. CSTR residence time for solvent fraction of 50%.

In reality, there is limited heat removal capacity for industrial reactors, so that heat evolution due to reaction is an additional source of autocatalytic behavior. Including these temperature effects produces even more interesting types of multiple steady state behavior. Figure 3 (taken from (147)) illustrates the types of steady state behavior which arise for methyl methacrylate polymerization when the solvent volume fraction is high enough that isothermal multiplicity is not possible and high conversion steady states are feasible operating conditions for the CSTR. Note that high conversions can be achieved at long residence times (large reactors) on the S-shaped multiplicity branch or alternatively at short residence times (small reactors) on the isolated branch. The isolated branch has the advantage that high conversion polymerization may be carried out with a reactor only 10% the size required for the same conversion on the continuous branch of steady states. However, these isolated steady states require special start-up procedures and reliable control systems to prevent extinction of the reaction. These model predictions have been verified by experiment for the polymerization of methyl methacrylate and for the polymerization of vinyl acetate (147). Figure 4 (taken from (147)) provides a detailed comparison of model prediction and experiment for vinyl acetate polymerization. Note that extinction is indeed observed at both ends of the isola.

Emulsion Polymerization in a CSTR. Emulsion polymerization is usually carried out isothermally in batch or continuous stirred tank reactors. Temperature control is much easier than for bulk or solution polymerization because the small (~.5 μm) polymer particles, which are the locus of reaction, are suspended in a continuous aqueous medium as shown in Figure 5. This complex, multiphase reactor also shows multiple steady states under isothermal conditions. Gerrens and coworkers at BASF seem to be the first to report these phenomena both computationally and experimentally. Figure 6 (taken from ref. (253)) plots the autocatalytic behavior of the reaction rate for styrene polymerization vs. monomer conversion in the reactor. The intersection of R_p with $\frac{(m_f-m)}{\tau}$ satisfies the material balance and determines the reactor steady states. These multiplicities occur for any polymerization having the Trommsdorf effect. For example, multiplicities for methyl methacrylate polymerization, found by Schork, et al. (204), are shown in Figure 7 where experimental steady state measurements are compared with model predictions. It is clear that the reactor designer would need to consider this isothermal multiplicity in selecting the desired reactor operating conditions.

Another exotic feature of emulsion polymerization in a CSTR is that sustained oscillations are frequently observed in

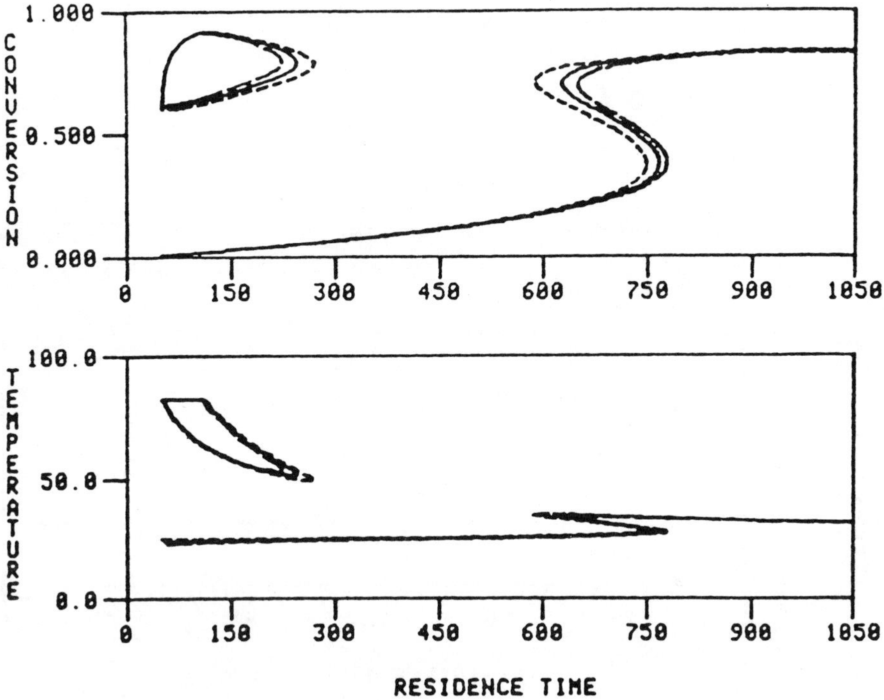

Figure 3. Effect of feed temperature on steady-state operation for MMA polymerization (147). Base case, T_f = 23 °C ——; T_f = 21 °C — — —; and T_f = 25 °C ----.

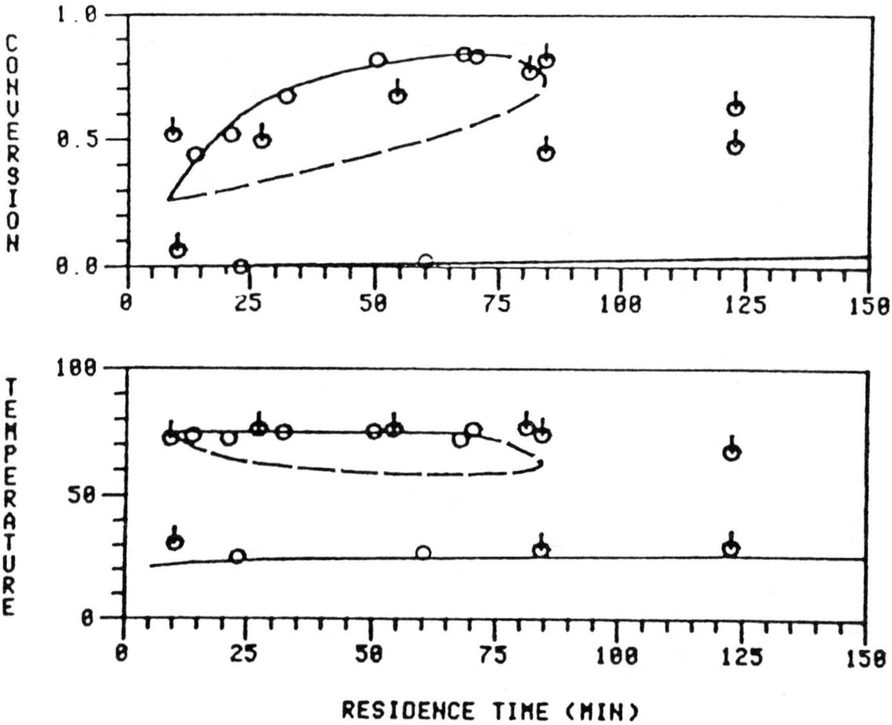

Figure 4. A comparison of the steady-state vinyl acetate model and experimental data (<u>147</u>). T_f = 20 °C; T_c = 25 °C; $_s$ = 0.50; I_f = 0.0417 mole/lit. $\overline{0}$ - stable steady state; and 0 - unstable transient.

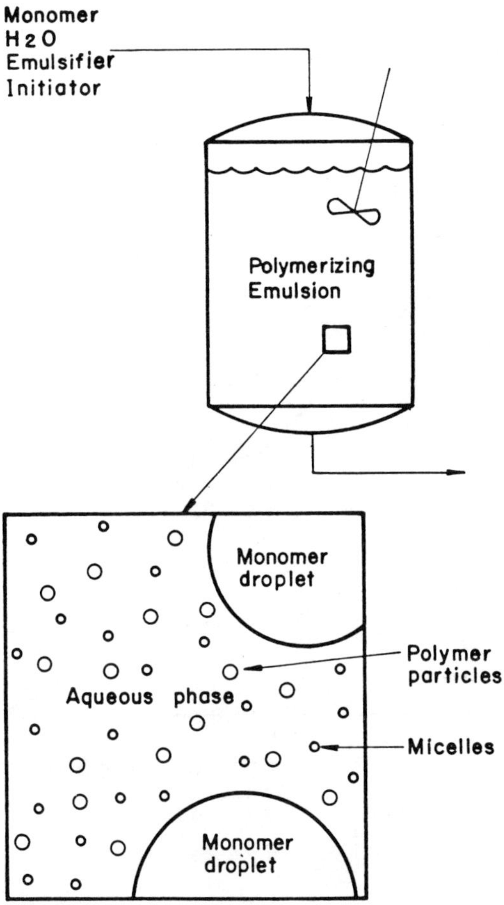

Figure 5. A continuous stirred tank emulstion polymerization reactor.

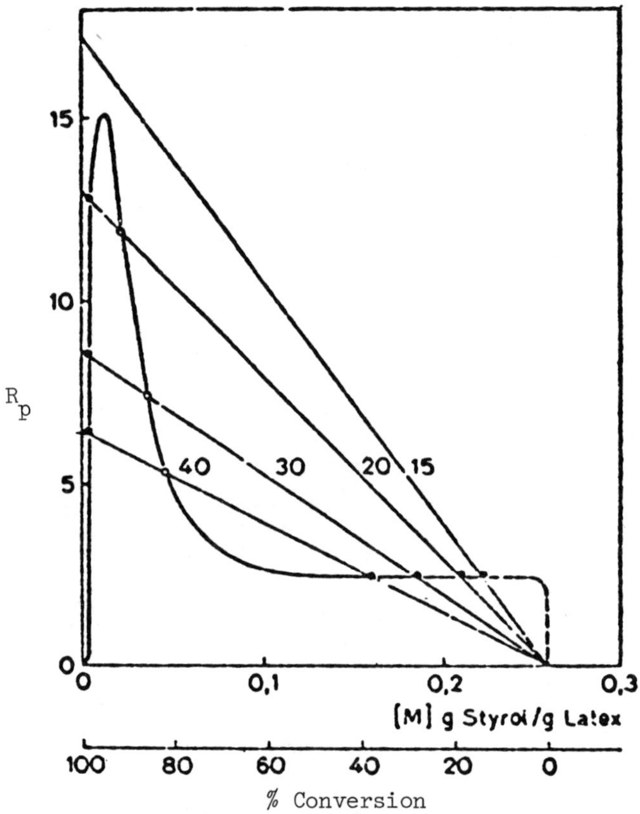

Figure 6. Multiple steady states in the emulsion polymerization of styrene in a CSTR. (Reproduced with permission from Ref. 253. Copyright 1971, Verlag Chemie, GMBH.)

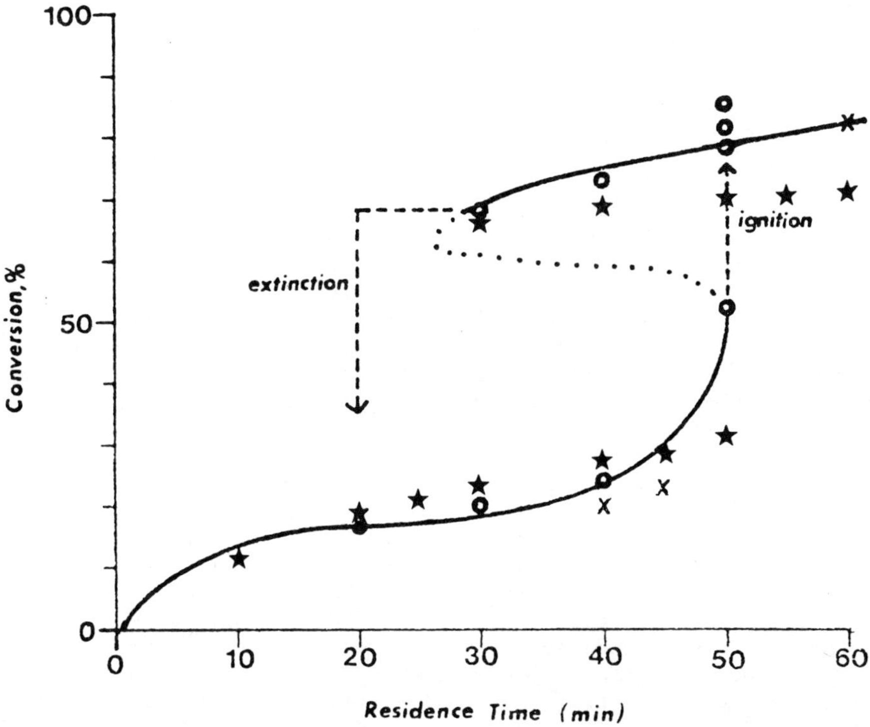

Figure 7. Isothermal multiplicity for the emulsion polymerization of methyl methacrylate in a CSTR (204). [S] = [I] = 0.03 moles/L-water; o,x - experimental steady states; and ★ - theoretical steady states.

industrial practice (cf. (20,253,254) for a discussion of this). Experimental results from our laboratory (201,202) shown in Figure 8 provide only one example of this troublesome problem. There is great interest in developing a quantitative understanding of the fundamental nature of these oscillations so that reactor design criteria for avoiding them can be developed. Recent developments by Rawlings (255) would seem to provide a first step in this direction.

Solid-Catalyzed Olefin Polymerization. A third area where extremely challenging polymerization reaction engineering problems arise is in transition-metal catalyzed polymerization of olefins such as ethylene, propylene and their copolymers. There is great controversy at present over issues such as (i) which factors control the molecular weight distribution and tacticity, (ii) what causes the decline in catalyst activity observed in these reactors, and (iii) what physical phenomena are occurring in the pores and on the catalyst surface. Unambiguous measurements are few, so theories abound. For example, olefin polymerization with a Ziegler-Natta type catalyst yields narrow molecular weight distributions when the system is homogeneous and very broad molecular weight distributions when the polymerization becomes heterogeneous. Has the nature of the catalyst changed so that some sites are producing short chains while other sites produce long chains? Or, has the increased heat and mass transfer resistance of the heterogeneous polymerization produced the great nonuniformity in polymer chain length?

To produce some answers to these and other questions requires a combination of mathematical modelling and diagnostic experimentation. There is some concensus that the physical picture inside the growing polymer particle is as follows. For heterogeneous catalytic systems polymerization occurs on the catalyst surface producing polymer chains which fill the pores and grow out from the active surface. The catalyst particle fractures into tiny fragments as the pores are ruptured by growing polymer. Monomer, which must diffuse to the catalyst surface to be catalytically inserted into the polymer chain, sees an increasing path for diffusion as polymerization progresses and the catalyst fragments become more and more encapsulated by polymer. Clearly this is a very complex physical situation and simple models are not adequate to represent all the important phenomena. Therefore a number of rather complex models for the catalyst particle have been developed and tested computationally with realistic parameters for heat and mass transfer as well as kinetics. Figure 9 illustrates some of these models while Figure 10 (taken from (256)) shows the predicted propylene monomer concentration profile to be expected in a polypropylene polymer particle over a 4 hour period in a slurry batch reactor. Figure 11 indicates the corresponding molecular weight distribution to be expected. As described in detail elsewhere (121,122, 124,256,257), these extensive modelling studies show that

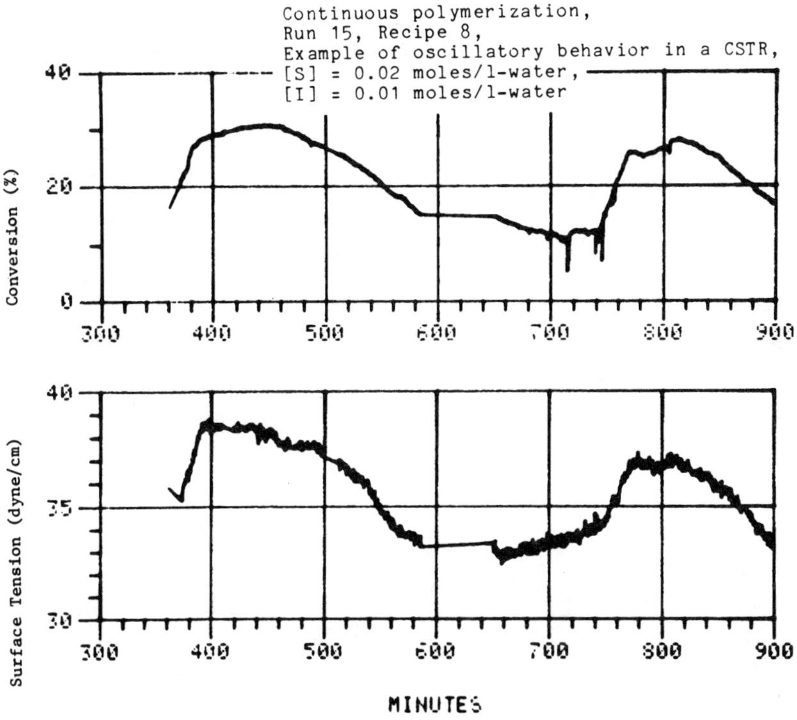

Figure 8. CSTR emulsion polymerization of methyl methacrylate. Residence time = 47 min., temperature = 40 °C (201). S = 0.02 moles/ -water; I = 0.01 moles/ -water.

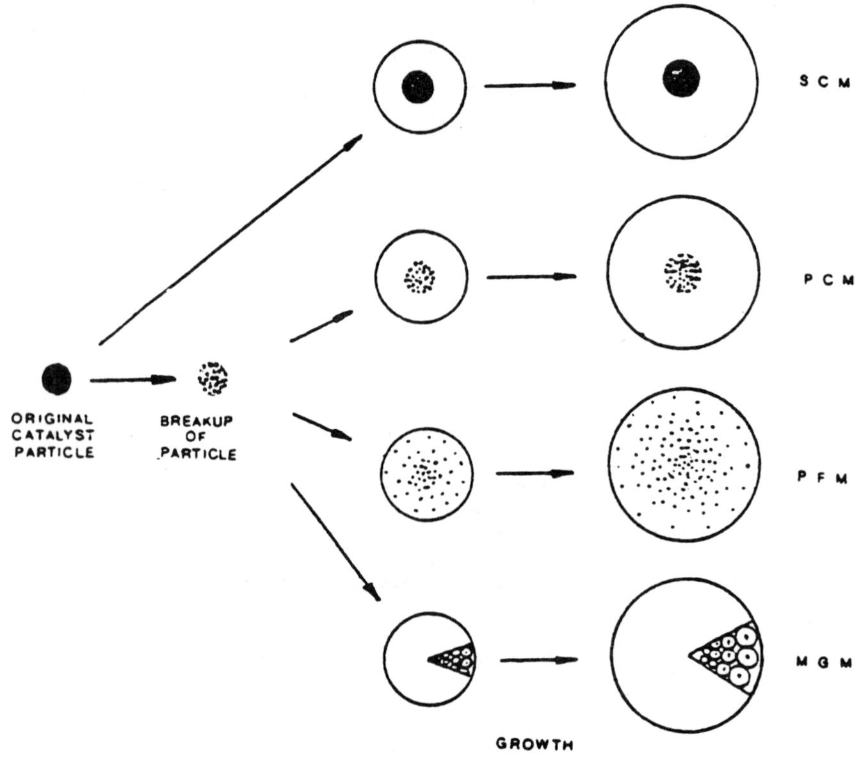

Figure 9. Schematic illustration of particle disintegration and the postulated morphological models; SCM (solid core model), PCM (polymeric core model), PFM (polymeric flow model), and MGM (multigrain model).

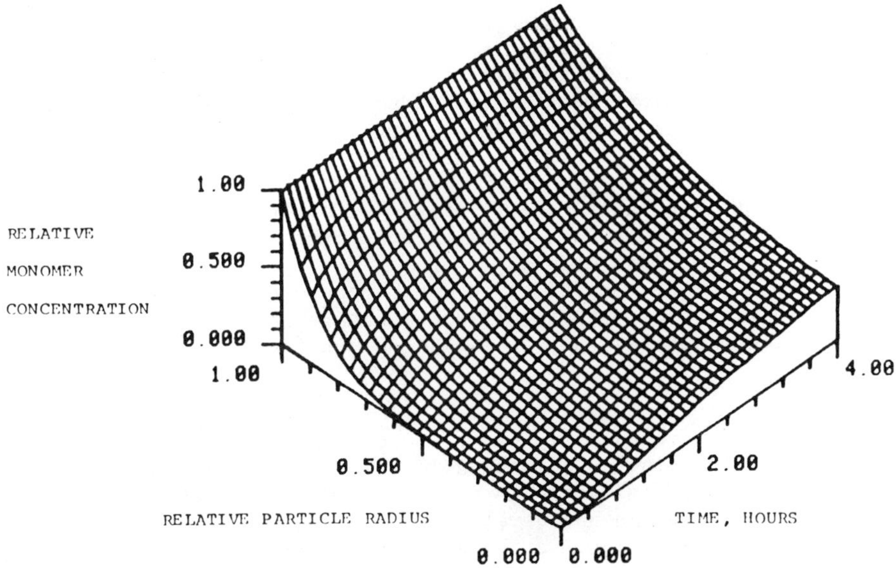

Figure 10. Propylene concentration profiles in the multigrain model as a function of radius and time for a high-activity catalyst.

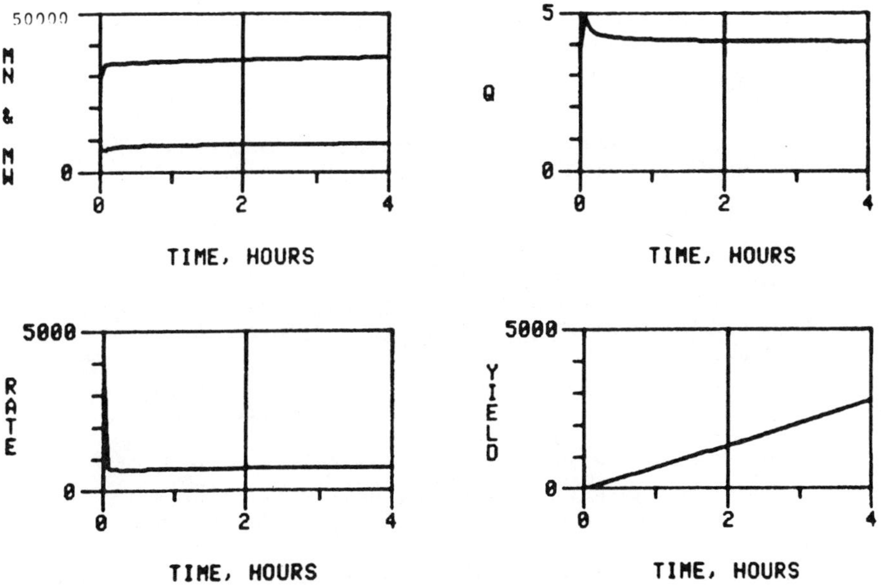

Figure 11. Multigrain model predictions for propylene polymerization in a semibatch reactor: $Q = M_w/M_n$, rate = gm polymer/gm cat.-hr, yield = gm polymer/gm-catalyst.

(i) mass transfer resistance can explain at least some of the observed broadening in MWD for ethylene and propylene polymerization.

(ii) Particle overheating can occur early in the polymerization, for small particles, and lead to sticking and agglomeration problems especially for gas phase processes.

(iii) Detailed quantitative matching of molecular weight distribution data requires even more complex mathematical models.

However, much remains to be done, both experimentally and theoretically, before there will be a comprehensive understanding of transition-metal catalyzed olefin polymerization.

Concluding Remarks

As may be evident from the few examples above, polymerization reaction engineering is a fascinating discipline, rich with diverse and challenging design problems. Currently there is considerable research and development activity in this area in industrial laboratories and there exists a high demand (even in these times) for Ph.D. graduates with polymerization reaction engineering expertise. However, as yet there is not a commensurate level of instruction and research in the university. Thus we in the university should include more from this area in our teaching. Similarly, government and industry could do more to provide the resources necessary to support university research and stimulate the number of Ph.D. graduates in this important area of chemical reaction engineering.

Acknowledgments

The author is indebted to the National Science Foundation and to the Mobil Oil Foundation for research support.

Literature Cited

1. C & E News, May 3, 1982, p 24.
2. Karol, F. G. U.S. Patents 4,011,382 and 4,646,647, 1979.
3. Keane, T. R. Proceedings 2nd ISCRE 1972, p A7 Elsevier, New York.
4. Ray, W. H. J. Macro. Sci.-Rev. Macro. Chem. 1972, C8, 1.
5. Keii, T. "Kinetics of Ziegler-Natta Polymerization", Chapman and Hall, London, 1972.
6. Albright, L. F. "Processes for Major Addition - Type Plastics and Their Monomers", McGraw-Hill, New York, 1974.
7. Min, K. W.; Ray, W. H. J. Macro. Sci.-Rev. Macro. Chem. 1974, C11, 177.
8. Gerrens, H. Proceedings 4th ISCRE 1976, p 585.
9. Bouton, T. C.; Chappelear, D. C. "Continuous Polymerization Reactors", AIChE Symposium Series 160, 1976.
10. Piirma, I.; Gardon, J. L. "Emulsion Polymerization", ACS Symp. Series 24, 1976.

11. Ray, W. H.; Laurence, R. L. "Polymerization Reaction Engineering", in "Chemical Reactor Theory", Lapidus, L.; Amundson, N. R., Ed., Prentice Hall, 1977, Chapter 9.
12. Poehlein, G. W.; Dougherty, D. J. J. Rubber Chem. Tech. 1977, 50, 601.
13. Schildknecht, C. E.; Skeist, I. "Polymerization Processes", Wiley, New York, 1977.
14. Henderson, J. N.; Bouton, T. C. "Polymerization Reactors and Processes", ACS Symposium Series 104, 1979.
15. Boor, J. "Ziegler-Natta Catalysis and Polymerization", Academic Press, New York, 1979.
16. Gerrens, H. "Polymerisationstechnik" in Ullmanns Encyklopädie der Technische Chemie 1980, 19, p 109.
17. Gerrens, H. Chem. Ing. Tech. 1980, 52, 477, translation in German Chemical Engineering 1981, 4, 1.
18. Gerrens, H. Chem. Tech. June, 1982, p 380, July 1982, p 434.
19. Odian, G. "Principles of Polymerization", Wiley, New York, 1981, 2nd Edition.
20. Ray, W. H. "Dynamic Behavior of Polymerization Reactors" in "Modelling of Chemical Reaction Systems", Ebert, K. H. et al., (Ed.), Springer Verlag, Berlin, 1981.
21. Bassett, D. R.; Hamielec, A. E. "Emulsion Polymers and Emulsion Polymerization", ACS Symp. Series 165, 1982.
22. Piirma, I. "Emulsion Polymerization", Academic Press, New York, 1982.
23. El-Aasser, M. S.; Vanderhoff, J. W. "Emulsion Polymerization of Vinyl Acetate", Appl. Sci. Publ., London, 1982.
24. Sebastian, D. H.; Biesenberger, J. A. "Polymerization Engineering", Wiley, New York, 1983, in press.
25. Quirk, R. P. Proceedings MMI Symposium on Transition Metal Catalyzed Polymerization (Gordon & Breach), 1983.
26. Dionisio, J. M.; O'Driscoll, K. F. J. Poly. Sci. Poly. Chem. 1980, 18, 3199.
27. Ito, K. Poly. J. 1980, 12, 499.
28. Tulig, T. J.; Tirrell, M. Macromol. 1981, 14, 1501.
29. Tulig, T. J.; Tirrell, M. Macromol. 1982, 15, 459.
30. Couchman, P. R. Poly. Eng. Sci. 1981, 21, 377.
31. Dvornić, P. R.; Jaćović, M. S. Poly. Eng. Sci. 1981, 21, 792.
32. Madruga, E. L.; San Roman, J.; del Puerto, M. A. Polymer 1981, 22, 951.
33. Marten, F. L.; Hamielec, A. E. J. Appl. Poly. Sci. 1982, 27, 489.
34. Gromov, V. F.; Galperina, N. I.; Osmanov, T. O.; Khomikovskii, P. M.; Abkin, A. D. Eur. Poly. J. 1980, 16, 529.
35. Yasukawa, T.; Murakami, K. Polymer 1980, 21, 1423.
36. Kulkarni, M. G.; Mashelkar, R. A. A.I.Ch.E. J. 1981, 27, 716.
37. Sundberg, D. C.; James, D. R. J. Poly. Sci. Poly. Chem. 1978, 16, 523.
38. Soh, S. K.; Sundberg, D. C. J. Poly. Sci. Poly. Chem. 1982, 20, 1299.

39. Soh, S. K.; Sundberg, D. C. J. Poly. Sci. Poly. Chem. 1982, 20, 1315.
40. Soh, S. K.; Sundberg, D. C. J. Poly. Sci. Poly. Chem. 1982, 20, 1331.
41. Soh, S. K.; Sundberg, D. C. J. Poly. Sci. Poly. Chem. 1982, 20, 1345.
42. Tulig, T. J.; Tirrell, M. Macromol. 1982, in press.
43. Tulig, T. J.; Tirrell, M. Macromol. 1981, 14, 1501.
44. Dubrovskii, S. A.; Kazanskii, K. S. Soviet Polymer Science 1981, p 775.
45. Ito, K.; Omi, M.; Ito, T. Poly. J. 1982, 14, 115.
46. Ito, K. Poly. J. 1982, 14, 101.
47. O'Driscoll, K. F. Pure & Appl. Chem. 1981, 53, 617.
48. Chapiro, A. Pure & Appl. Chem. 1981, 53, 643.
49. Kodaira, Y.; Harwood, H. J. 182nd ACS Meeting, N.Y. 1981, 305.
50. Johnson, A. F.; Khaligh, B.; Ramsay, J. Int. J. Modelling Simul. 1981, 1, 313.
51. Johnson, A. F.; Khaligh, B.; Ramsay, J. ACS Meeting N.Y. 1981, 299.
52. Tirrell, M., Gromley, K. Chem. Eng. Sci. 1981, 36, 367.
53. Bejger, T. P.; Tirrell, M. V.; Stephanopoulos, G. A.I.Ch.E. Meeting New Orleans, 1981.
54. Tsoukas, A.; Tirrell, M. Chem. Eng. Sci., in press.
55. Sebastian, D. H.; Biesenberger, J. A. Poly. Eng. Sci. 1979, 19, 190.
56. Motoc, I.; Vancea, R. J. Poly. Sci. Poly. Chem. 1980, 18, 1559.
57. Lopez-Serrano, F.; Castro, J. M.; Macosko, C. W.; Tirrell, M. Polymer 1980, 21, 263.
58. Ivin, K. J.; Pitchumani, S.; Reddy, C. R.; Rajadurai, S. Eur. Poly. J. 1981, 17, 341.
59. Probst, J.; Kolb, G. Angew. Makromol. Chem. 1981, 98, 195.
60. Kuo, J.-F.; Chen, C.-Y. J. Appl. Poly. Sci. 1981, 26, 1117.
61. Kumar, A.; Gupta, S. K.; Saraf, R. Polymer 1980, 21, 1323.
62. Katuscak, S.; Tomas, M.; Schiessl, O. J. Appl. Poly. Sci. 1981, 26, 381.
63. Tai, K.; Arai, Y.; Teranishi, H.; Tagawa, T. J. Appl. Poly. Sci. 1980, 25, 1789.
64. Gandhi, K. S.; Babu, S. V. Macromol. 1980, 13, 791.
65. Gupta, S. K.; Kumar, A.; Saraf, R. J. Appl. Poly. Sci. 1980, 25, 1049.
66. Sebenda, J. Prog. Poly. Sci. 1978, 6, 123.
67. Kumar, A.; Gupta, S. K.; Phukan, U. K. Poly. Eng. Sci. 1981, 21, 1218.
68. Herzfeld, J.; Briehl, R. W. Macromol. 1981, 14, 1209.
69. Kumar, A.; Kuruville, S.; Raman, A. R.; Gupta, S. K. Polymer 1981, 22, 389.
70. Gupta, S. K.; Kumar, A.; Tandon, P.; Naik, C. D. Polymer 1981, 22, 481.

71. Heikens, D. Polymer 1981, 22, 1758.
72. Kumar, A.; Sukthankar, V. K.; Vaz, C. P.; Gupta, S. K. Polymer, in press.
73. Kumar, A.; Madan, S.; Shah, N. G.; Gupta, S. K. Polymer, in press.
74. Aneja, A. P. Poly. Eng. Rev., in press.
75. Gajewski, H.; Zacharias, K. Acta Polymerica 1981, 32, 57.
76. Rafler, G.; Bonatz, E.; Reinisch, G.; Gajewski, H.; Zacharias, K. Acta Poly. 1980, 31, 570.
77. Rafler, G.; Blaesche, J.; Gajewski, H.; Zacharias, K. Acta Poly. 1980, 31, 633.
78. Rafler, G.; Bonatz, E.; Gajewski, H.; Zacharias, K. Acta Poly. 1980, 31, 684.
79. Rafler, G.; Bonatz, E.; Reinisch, G.; Gajewski, H.; Zacharias, K. Acta Poly. 1980, 31, 732.
80. Gajewski, H.; Zacharias, K. Z.A.M.M. 1980, 60, 357.
81. Frunze, T. M.; Shleifman, R. B.; Belavtseva, E. M.; Genin, Ya. V.; Volkova, T. V.; Kotel'nikov, V. A.; Radchenki, L. G.; Davtyan, S. P.; Kuraschev, V. V.; Tsvankin, D. Ya. J. Poly. Sci. Poly. Phys. 1980, 18, 1523.
82. Ishida, S.-I.; Tsutsumi, Y.; Kaneko, K. J. Poly. Sci. Poly. Chem. 1981, 19, 1609.
83. Kumar, A.; Rajora, P.; Agarwalla, N. L.; Gupta, S. K. Polymer 1982, 23, 222.
84. Kumar, A.; Phukan, U. K.; Kulshreshtha, A. K.; Gupta, S. K. Polymer 1982, 23, 215.
85. Borrajo, J.; Aranguren, M. I.; Williams, R. J. J. Polymer 1982, 23, 263.
86. Frontini, P. M.; Cuadrado, T. R.; Williams, R. J. J. Polymer 1982, 23, 267.
87. Pomakis, I.; Simitzis, I. Angw. Makromol. Chem. 1981, 99, 145.
88. Gupta, S. K.; Naik, C. D.; Tandon, P.; Kumar, A. J. Appl. Poly. Sci. 1981, 26, 2153.
89. Ravindranath, K.; Mashelkar, R. A. J. Appl. Poly. Sci. 1981, 26, 3179.
90. Durand, D.; Bruneau, C.-M. Polymer 1982, 23, 69.
91. Kumar, A.; Agarwal, R. K.; Gupta, S. K. J. Appl. Poly. Sci. 1982, 27, 1759.
92. Ravindranath, K.; Mashelkar, R. A. J. Appl. Poly. Sci. 1982, 27, 471.
93. Lin, C.-C.; Chiu, W.-Y. J. Appl. Poly. Sci. 1982, 27, 63.
94. Gupta, A.; Gandhi, K. S. J. Appl. Poly. Sci. 1982, 27, 1099.
95. Bohm, L. L. Angw. Makromol. Chem. 1980, 89, 1.
96. Michael, R.; Reichert, K. H. Can. J. Ch.E. 1981, 59, 602.
97. Keszler, B.; Gröbler, A.; Takacs, E.; Simon, A. Polymer 1981, 22, 818.
98. Munoz-Escalona, A.; Martinez, C.; Hidalgo, J. Polymer 1981, 22, 1118.

99. Munoz-Escalona, A.; Villamizar, C.; Frias, P. in "Structure-Property Relationships of Polymeric Solid", Plenum, NY, 1982.
100. Munoz-Escalona, A.; Parada, A. J. Crystal Growth 1980, 48, 250.
101. Franz, H.; Meyer, H.; Reichert, K.-H. Polymer 1981, 22, 226.
102. Reichert, K. H. Angw. Makromol. Chem. 1981, 94, 1.
103. Michale, R.; Reichert, K. H. Can. J. Chem. Eng. 1981, 59, 602.
104. Hogan, J. P.; Norwood, D. D.; Ayres, C. A. J. Appl. Poly. Sci. Appl. Poly. Sci. Symp. 1981, 36, 49.
105. Vermel', Ye. Ye.; Zakharov, V. A.; Bukatova, Z. K.; Shkurina, G. P.; Yechevskaya, L. G.; Moroz, E. M.; Sudakova, S. V. Poly. Sci. USSR 1980, 22, 23.
106. Pino, P.; Mülhaupt, R. Angew. Chem. Int. Ed. Engl. 1980, 19, 857.
107. Reichert, K. H. Proc. MMI Symp. Transition Metal Catal. Poly. 1981
108. Doi, Y.; Ueki, S.; Tamura, S.; Nagahara, S.; Keii, T. Polymer 1982, 23, 258.
109. McDaniel, M. P. J. Poly. Sci. Poly. Chem. 1981, 19, 1967.
110. Sailors, H. R.; Hogan, J. P. J. Macromol. Sci-Chem. 1981, A15, 1377.
111. Zakharov, V. A. Kinet. Katal. 1980, 21, 892.
112. Likholobov, V. A.; Ermakov, Yu. I. Kinet. Katal. 1980, 21, 904.
113. Hoff, R. E.; Pullukat, T. J.; Shida, M. J. Appl. Poly. Sci. 1981, 26, 2927.
114. Vermel', E. E.; Zakharov, V. A.; Fenelonov, V. B.; Zaikovskii, V. I.; Zagrafskaya, R. V.; Moroz, E. M.; Bukatov, G. D. Kinet. Katal. 1981, 22, 480.
115. Brockmeier, N. F. MMI Symp. Transition Metal Catal. Poly. 1081.
116. Brockmeier, N. F.; Lin, C. H. A.I.Ch.E. Meeting Houston 1979.
117. Doi, Y.; Nishimura, Y.; Keii, T. Polymer 1981, 22, 469.
118. Tsutsui, T.; Tanaka, R. Polymer 1980, 21, 1351.
119. Doi, Y.; Ueki, S.; Keii, T. Macromol. 1979, 12, 814.
120. Doi, Y.; Takada, M.; Keii, T. Bull. Chem. Soc. Japan 1979, 52, 1902.
121. Nagel, E. J.; Kirillov, V. A.; Ray, W. H. I&EC Product Res. & Dev. 1980, 19, 372.
122. Taylor, T. W.; Choi, K. Y.; Yuan, H.-G.; Ray, W. H. Proc. Int. Symp. Transition Metal Catal. Poly. (Gordon & Breach), 1982.
123. Yuan, H. G.; Taylor, T. W., Choi, K. Y.; Ray, W. H. J Appl. Poly. Sci. 1982, 27, 1691.
124. Choi, K. Y.; Taylor, T. W.; Ray, W. H. Proc. 1982 IUPAC Macro., Aug. 1982.
125. Yamamoto, K. J. Macromol. Sci.-Chem. 1982, A17, 415.

126. Seidl, H.; Luft, G. J. Macromol. Schi.-Chem. 1981, A15, 1.
127. Mattice, W. L.; Stehling, F. C. Macromol. 1981, 14, 1479.
128. Romanini, D.; Savadori, A. Polymer 1980, 21, 1092.
129. Gianotti, G.; Cicuta, A. Polymer 1980, 21, 1087.
130. Takahashi, T.; Ehrlich, P. 1982, paper in press.
131. Bohdanecky, M. Makromol. Chem. 1980, 181, 1321.
132. Bersted, B. H.; Slee, J. D.; Richter, C. A. J. Appl. Poly. Sci. 1981, 26, 1001.
133. Stoiljkovich, D.; Jovanovich, S. J. Poly. Sci. Poly. Chem. 1981, 19, 741.
134. Stoiljkovich, D.; Jovanovich, S. Makromol. Chem. 1981, 182, 2811.
135. Yamamoto, K. Preprints Atlanta ACS Meeting, April 1981.
136. Brooks, B. W. Chem. Eng. Sci. 1981, 36, 589.
137. Chen, H. T.; Chartier, P. A.; Setthachayanon, S. Poly. Eng. Sci. 1980, 20, 1197.
138. Chen, H. T.; Kuan, C. N.; Setthachayanon, S.; Chartier, P. A. A.I.Ch.E. J. 1980, 26, 672.
139. Chen, H. T.; Kuan, C. N.; Lin, D. J. A.I.Ch.E. J. 1982, 28, 215.
140. Stolzenberg, K.; Kirchner, K. Angw. Makromol. Chem. 1981, 95, 185.
141. Baade, W.; Reichert, K.-H. Chem.-Ing.-Tech. 1981, 53, 472.
142. Umoh, N. F.; Harriott, P.; Hughes, R. Chem. Eng. J. 1981, 21, 85.
143. Meira, G. R. J. Macromol. Sci.-Rev. Macromol. Chem. 1981, C20, 207.
144. Nassar, R.; Too, J. R.; Fan, L. T. J. Appl. Poly. Sci. 1981, 26, 3745.
145. Schmidt, A. D.; Ray, W. H. Chem. Eng. Sci. 1981, 36, 1401.
146. Hamer, J. W.; Akramov, T. A.; Ray, W. H. Chem. Eng. Sci. 1981, 36, 1897.
147. Schmidt, A. D.; Clinch, A. B.; Ray, W. H. Chem. Eng. Sci. 1983 (submitted for publication).
148. Siadat, B.; Malone, M.; Middleman, S. Poly. Eng. Sci. 1979, 19, 787.
149. Mihail, R.; Straja, S. Chem. Eng. Sci. 1981, 36, 1263.
150. Quevedo, J.; White, D. H. 2nd World Congress Chem. Eng. 1981, 535.
151. Janssen, L. P. B. M.; Schaart, B. J. 2nd World Congress Chem. Eng. 1981, 543.
152. Lindt, J. T. 2nd World Congress Chem. Eng. 1981, 539.
153. Kemblowski, Z.; Sek, J. Poly. Eng. Sci. 1981, 21, 1194.
154. Zhirkov, P. V.; Bostandzhiyan, S. A.; Boyarchenko, V. I. Teoret. Osnovy Khim. Tekh. 1980, 14, 702.
155. Vol'ter, B. V.; Sofiev, A. E.; Savel'ev, A. M.; Murav'eva, Tumarkin, N. Ya.; Faitel'son, L. A.; Briedis, I. P. Teoret. Osnovy Khim. Tekh. 1980, 14, 894.
156. Chella, R.; Ottino, J. M. Proc. 7th ISCRE 1982.
157. Weiss, W.; Böhme, P. Chem. Techn. 1981, 33, 10.

158. Joshi, S. D.; Bergles, A. E. A.I.Ch.E. J. 1981, 27, 872.
159. Joosten, G. E. H.; Hoogstraten, H. W.; Cuwerkerk, C. Ind. Eng. Chem. Proc. Des. Dev. 1981, 20, 177.
160. Lynch, D.; Kiparissides, C. J. Appl. Poly. Sci. 1981, 26, 1283.
161. Panke, D. Makromol. Chem. 1981, 182, 3207.
162. Budtov., V. P.; Ponomareva, Ye. L.; Belyayev, V. M. Vysokomol. soyed. 1980, A22, 2152.
163. Saxena, A. K.; Nigam, K. D. P. J. Appl. Poly. Sci. 1981, 26, 3475.
164. Singh, D.; Nigam, K. D. P. J. Appl. Poly. Sci. 1981, 26, 785.
165. Agur, E. E.; Vlachopoulos, J. J. Appl. Poly. Sci. 1981, 26, 765.
166. Hamer, J. W.; Ray, W. H. 1981 Annual A.I.Ch.E. Meeting New Orleans.
167. Richter, E. B.; Macosko, C. W. Poly. Eng. Sci. 1980, 20, 921.
168. Miller, D. R.; Valles, E. M.; Macosko, C. W. Poly. Eng. Sci. 1979, 19, 272.
169. Lee, L. J.; Ottino, J. M.; Ranz, W. E.; Macosko, C. W. Poly. Eng. Sci. 1980, 20, 868.
170. Domine, J. D.; Gogos, C. G. Poly. Eng. Sci. 1980, 20, 847.
171. Biesenberger, J. A.; Gogos, C. G. Poly. Eng. Sci. 1980, 20, 838.
172. Steinle, E. C.; Critchfield, F. E.; Castro, J. M.; Macosko, C. W. J. Appl. Poly. Sci. 1980, 25, 2317.
173. Throne, J. L.; Gianchandani, J. Poly. Eng. Sci. 1980, 20, 899.
174. Castro, J. M.; Macosko, C. W. A.I.Ch.E. J. 1982, 28, 250.
175. Wiley, S. J. 2nd World Congress Chem. Eng. 1981, 45; Stevenson, J. F. 2nd World Congress Chem. Eng. 1981, 447.
176. Hedrick, R. W.; Gabbert, J. D. 2nd World Congress Chem. Eng. 1981, 461.
177. Throne, J. L.; Gianchandani, J.; Progelhof, R. C. 2nd World Congress Chem. Eng. 1981, 465.
178. Kamal, M. R.; Tan, V.; Sid, S. 2nd World Congress Chem. Eng. 1981, 471.
179. Ottino, J. M.; Chella, R. 2nd World Congress Chem. Eng. 1981, 475.
180. Lambla, M.; Druz, J. 2nd World Congress Chem. Eng. 1981, 504.
181. Gonzalez, V. M.; Castro, J. M.; Macosko, C. W. 2nd World Congress Chem. Eng. 1981, 519.
182. Manzione, L. T. 2nd World Congress Chem. Eng. 1981, 526.
183. Tirrell, M.; Lee, L. J.; Macosko, C. W. ACS Symp. Series No. 104 1979, 149.
184. Castro, J. M.; Lopez-Serrano, F.; Camargo, R. E.; Macosko, C. W.; Tirrell, M. J. Appl. Poly. Sci. 1981, 26, 2067.

185. Kovenklioglu, S. Poly. Eng. Sci. 1980, 20, 816.
186. Chen, S.-A.; Huang, N.-W. Chem. Eng. Sci. 1981, 36, 1295.
187. Stolin, A. M.; Merzhanov, A. G.; Malkin, A. Y. Poly. Eng. Sci. 1979, 19, 1065.
188. Timm, D. C.; Gilbert, R. E.; Ko, T. T.; Simmons, M. R. 182nd ACS Meeting N.Y. 1981, 194.
189. Cutter, L. A.; Drexler, T. D. 182nd ACS Meeting N.Y. 1981, 200.
190. Harwood, H. J.; Dworak, A.; Nyeu, T. K. L.; Tong, S.-N. 182nd ACS Meeting N.Y. 1981, 220.
191. Lodge, A. S. Proc. 1982 IUPAC Conf., Aug. 1982.
192. Hulme, J. M.; Thibodeau, W. E. ISA 1980, 39.
193. Trathnigg, B. Makromol. Chem. 1980, 181, 1979.
194. Gusika, P. L.; Zhizhin, V. G. PMM USSR 1981, 44, 495.
195. Blanks, R. F.; Meyer, J. A.; Grulke, E. A. Poly. Eng. Sci. 1981, 21, 1055.
196. Wu, G. Z. A.; Denton, L. A.; Laurence, R. L. Poly. Eng. Sci. 1982, 22, 1.
197. Meira, G. R.; Johnson, A. F. Poly. Eng. Sci. 1981, 21, 415.
198. Rao, B. B.; Mhaskar, R. D. Polymer 1981, 22, 1593.
199. Hwu, M. C.; Foster, R. D. CEP July 1982, 62.
200. Zamani, H.; Daroux, M.; Gregge, J. L.; Bordet, J. 2nd World Congress Chem. Eng. 1981, 535.
201. Schork, F. J.; Ray, W. H. ACS Sym. Series 1981, 165, 505.
202. Schork, F. J.; Ray, W. H. J. Appl. Poly. Sci. 1983, 28, in press.
203. Schork, F. J.; Chu, G. C.; Ray, W. H. 1980 Annual A.I.Ch.E. Meeting Chicago.
204. Schork, F. J.; Rawlings, J. B.; Ray, W. H. 1981 Annual A.I.Ch.E. Meeting New Orleans.
205. Kiparissides, C.; MacGregor, J. F.; Hamielec, A. E. Can. J. Chem. Eng. 1980, 58, 48, 56, 65.
206. Allsopp, M. W. Pure & Appl. Chem. 1981, 53, 449.
207. Yang, C. M.; Hamann, H. C. J. Appl. Poly. Sci. 1980, 25, 2555.
208. Hatate, Y.; Ikeura, T.; Shinonome, M.; Kondo, K.; Nakashio, F. J. Chem. Eng. Japan 1981, 14, 38.
209. Lin, C.-C.; Wang, Y.-F. J. Appl. Poly. Sci. 1981, 26, 3909.
210. Schröder, R.; Piotrowski, B. Chem.-Ing.-Tech. 1981, 53, 748.
211. Cameron, J. B.; Lundeen, A. J.; McCulley, J. H., Jr.; Schwab, P. A. 181st ACS Meeting Atlanta 1981.
212. Moritz, H.-U.; Reichert, K. H. Chem.-Ing. Tech. 1981, 53, 386.
213. Langner, F.; Reichert, K. H. Chem.-Ing.-Tech. 1981, 53, 747.
214. Langner, F.; Moritz, H.-U.; Reichert, K. H. Ger. Chem. Eng. 1979, 2, 329.
215. Moritz, H.-U.; Langner, F.; Reichert, K. H. Ger. Chem. Eng. 1979, 2, 112.

216. Baade, W.; Reichert, K. H. Chem.-Ing.-Tech. 1981, 53, 472.
217. Langner, F.; Moritz, H.-U.; Reichert, K. H. Chem. Eng. Sci. 1980, 35, 519.
218. Brooks, B. W. J.C.S. Faraday I 1980, 76, 1599.
219. Guillot, J. Makromol. Chem. Rapid Commun. 1980, 1, 697.
220. Kiparissides, C.; MacGregor, J. F.; Hamielec, A. E. A.I.Ch.E. J. 1981, 27, 13.
221. Nomura, M.; Harada, M. J. Appl. Poly. Sci. 1981, 26, 17.
222. Rance, D. G.; Zichy, E. L. Pure & Appl. Chem. 1981, 53, 377.
223. Tseng, C. M.; El-Aasser, M. S.; Vanderhoff, J. W. ACS Meeting N.Y. 1981.
224. Ugelstad, J.; Mork, P. C.; Hansen, F. K. Pure & Appl. Chem. 1981, 53, 323.
225. Oldshue, J. Y.; Mechler, D. O.; Grinnell, D. W. CEP May 1982, 68.
226. Bataille, P.; Van, B. T.; Pham, Q. B. J. Poly. Sci. Poly. Chem. 1982, 20, 795.
227. Cauley, D. A.; Thompson, R. W. J. Appl. Poly. Sci. 1982, 27, 363.
228. Piirma, I.; Chang, M. J. Poly. Sci. Poly. Chem. 1982, 20, 489.
229. Almog, Y.; Levy, M. J. Poly. Sci. Poly. Chem. 1982, 20, 417.
230. Brooks, B. W.; Makanjuola, B. O. Polymer 1982, 23, 77.
231. Ham, G. E. J. Macromol. Sci.-Chem. 1982, A17, 369.
232. James, D. R.; Sundberg, D. C. J. Poly. Sci. Poly. Chem. 1980, 18, 903.
233. Sundberg, D. C.; Hsieh, J. Y.; Soh, S. K.; Baldus, R. F. ACS Symp. Ser. No. 165 1981, 327.
234. Dautzenberg, H.; Goebel, H.-H.; Koch, H.; Jaeger, W.; Linow, K.-J.; Purz, H. J.; Reinisch, G.; Schneider, H. J.; Richter, I.; Wolf, J. Acta Poly. 1980, 31, 335.
235. Friedman, E. M. J. Poly. Sci. Poly. Phys. 1980, 18, 1771.
236. Piirma, I.; Chen, S.-R. J. Coll. Interface Sci. 1980, 74, 90.
237. Min, K. W.; Gostin, H. I. Ind. Eng. Chem. Prod. Res. Dev. 1979, 18, 272.
238. Kiparissides, C.; Ponnuswamy, S. R. Chem. Eng. Commun. 1981, 10, 283.
239. Nomura, M.; Harada, M. ACS Symp. Ser. No. 165 1981, 121.
240. Nomura, M.; Kubo, M.; Kujita, K. Mem. Fac. Eng. Fukui Univ. 1981, 29, 167.
241. Nomura, M.; Yamamoto, K.; Horie, I.; Fujita, K. J. Appl. Poly. Sci., in press.
242. Kawaguchi, H.; Sugi, Y.; Ohtsuka, Y. ACS Symp. Ser. No. 165 1981, 145.
243. Bassett, D. R.; Hoy, K. L. ACS Symp. Ser. No. 165 1981, 371.

244. Johnston, J. E.; Bassett, D. R.; MacRury, T. B. ACS Symp. Ser. No. 165 1981, 389.
245. Guyot, A.; Guillot, J.; Pichot, C.; Guerrero, L. R. ACS Symp. Ser. No. 165 1981, 415.
246. Patterson, I.; Rollin, A. L.; Estaque, L.; Bataille, P. A.I.Ch.E. Meeting Chicago 1980.
247. Guillot, J. Makromol. Chem. Rapid Commun. 1980, 1, 697.
248. Ohtsuka, Y.; Kawaguchi, H.; Sugi, Y. J. Appl. Poly. Sci. 1981, 26, 1637.
249. Chonde, Y.; Krieger, I. M. J. Appl. Poly. Sci. 1981, 26, 1819.
250. Snuparek, J., Jr.; Kaspar, K. J. Appl. Poly. Sci. 1981, 26, 4081.
251. Lin, C.-C.; Ku, H.-C.; Chiu, W.-Y. J. Appl. Poly. Sci. 1981, 26, 1327.
252. Ham, G. E. J. Macromol. Sci.-Chem. 1982, A17, 369.
253. Gerrens, H.; Kuchner, K.; Ley, G. Chem. Ing. Tech. 1971, 43, 693-98.
254. Kirillov, V. A.; Ray, W. H. Chem. Eng. Sci. 1978, 33, 1499.
255. Rawlings, J. B.; Ray, W. H. 1982 Annual A.I.Ch.E. Meeting Los Angeles.
256. Taylor, T. W.; Ray, W. H. 1982 Annual A.I.Ch.E. Meeting Los Angeles.
257. Choi, K. Y.; Taylor, T. W.; Ray, W. H. J. Appl. Poly. Sci. (submitted for publication).

RECEIVED April 26, 1983

6

Mixing in Chemical Reactors

JACQUES VILLERMAUX

Institut National Polytechnique de Lorraine, Laboratoire des Sciences du Génie Chimique, CNRS-ENSIC, Nancy, France

Recent progress in the understanding and modelling of mixing phenomena in chemical reactors is reviewed. The following items are discussed: characterization of the degree of mixing, Eulerian approach to turbulence theory with application to modelling combustion, macromixing and residence time distributions, phenomenological models for mixing earliness in the age space, mechanism of micromixing in physical space and segregation phenomena at the microscopic level, mixing in stirred tanks and static mixers. An example is presented of the application to continuous free radical polymerization. It is concluded that a comprehensive and predictive theory for micromixing should not be sought through turbulence theory alone, but rather in phenomenological interaction models, whose parameters could be interpreted on a fundamental basis by this theory.

Contacting reactants and subsequent mixing of reacting species is one of the major features controlling the behaviour of chemical reactors. In spite of recent advances, a unified theory providing the engineer with general rules applicable to any case of competition between mixing and chemical reaction is still lacking. However, directions in which such a theory should be sought are now in view. The aim of this chapter is to review recent and significant contributions which may lead to such a general treatment.

Of course, the field of mixing in chemical engineering is immense. Therefore, this review will be restricted, with perhaps a few exceptions, to those phenomena where mixing and chemical reactions are closely linked. The purely physical aspects of mixing won't be considered, although this important operation also poses unsolved problems to industry.

That mixing in chemical reactors is a subject of very active research is attested by the recent publication of several review

0097–6156/83/0226–0135$13.80/0

papers, which appeared a few years ago or while the present was in preparation (1-6).

Instead of duplicating these reviews and rather than presenting an exhaustive literature survey, I have chosen to emphasize the points which seem the most critical and controversial, or those where significant progress has been made in recent years, and to give personal views on these subjects. In doing so, I accept the risk that this review may appear partial or incomplete to those authors whose (sometimes excellent) papers will be left out.

That mixing in chemical reactors is also a problem of vital importance to industry appears in some, too rare, papers describing the effect of agitation on yield in pilot plant reactors (7) or the difficulties of conserving a good selectivity when scaling up a mixing device to the commercial scale (8). It is a pity that proprietary requirements restrain the publication of such industrial case studies, whose analysis would be of the greatest interest to theoricians of mixing. But this is a general problem in Chemical Reaction Engineering. Nevertheless, it is encouraging to see industrial participants attending meetings and giving their opinion about the relevance of research topics and about what should be done to improve our knowledge in the field (9, 10). This should prevent "too many academics pursuing too many non-problems" (11).

The last point that should be outlined in this introduction is that mixing in chemical reactors really constitutes an original chapter of Chemical Reaction Engineering. Since the pioneering work of Danckwerts (12), Zwietering (13) and others, it is clear that chemical reaction engineering constitutes a scientific discipline with its own methods. The RTD and the segregation concepts for instance, have completely renewed the description of the behaviour of reacting mixtures in reactors. I am convinced that future progress can be expected in pursuing novel concepts rather than in more sophisticated combinations of chemical kinetics and fluid mechanics. Of course this doesn't mean that we must neglect the contribution of the latter discipline, as will be seen below.

Definition and characterization of the degree of mixing

A definition of mixing is proposed in many papers and textbooks (14, 15, 16, 2), and this point will not be discussed here. Several criteria are used for a quantitative characterization of the "quality of mixing". These have been recalled by Hiby (17). In a non-uniform mixture, let p(C) be the local concentration distribution of a species. p(C) may be characterized by its mean $\overline{C}$, its variance σ^2 and the average value $\overline{\Delta C}$ of the deviation $\Delta C = |C - \overline{C}|$ from the mean. From these quantities, several criteria may be defined (17), namely $\delta = \overline{\Delta C}/\overline{C}$, $\delta_{max} = \Delta C_{max}/\overline{C}$ and $\delta_\sigma = \sigma/\overline{C}$. Another criterium is $\Delta = \overline{\Delta C}/\overline{\Delta C}_o$, where $\overline{\Delta C}_o$ is the value of $\overline{\Delta C}$ before mixing (or at the reactor inlet). Starting from two streams of (reduced) concentration 0 and 1, then $\overline{\Delta C}_o = 2\,\overline{C}(1 - \overline{C})$. The

Intensity of Segregation defined by Danckwerts (14) is $I_s = \sigma^2/\sigma_o^2 = \sigma^2/\overline{C}(1 - \overline{C})$, but $\sqrt{I_s} = \sigma/\sigma_o$ has also been used. If two species A and B are mixed, the definition of I_s becomes (2) $I_s = -\overline{c_A c_B}/C_{Ao}.C_{Bo}$. Relationships between all these indices are obvious. Their common property is that the smaller they are, the better the quality of the mixture. The corresponding degrees of homogeneity may also be defined as $1 - \delta$, $1 - \delta_\sigma$, $1 - \sqrt{I_s}$, etc...

Mixing times are the times required for δ, or Δ, or I_s to fall from their initial value (before mixing), down to some prescribed small value (for instance 0.05 or 0.01). We shall see that in Lagrangian coordinates, $- \sigma^2/(d\sigma^2/dt)$ is also a mixing time. The estimation of mixing times in stirred tanks will be discussed in a further Section.

The scale of segregation gives information on the size of segregated regions. Let c(x) and c(x+r) be the fluctuating concentrations at points x and x+r ($c = C - \overline{C}$), then the autocorrelation function of concentration is (2): $g_s(r) = \overline{c(x)\ c(x+r)}/\sigma^2$.

The scale of segregation is then defined as :

$$L_s = \int_0^\infty g_s(r)\ dr$$

In turbulence theory, this is also known as the concentration macroscale, which plays an important role in the interpretation of micromixing phenomena.

A first remark about these mixing indices is that a certain confusion prevails in the literature owing to the multiplicity of definitions. It would be desirable to adopt one single measure for the degree of mixing (and/or homogeneity), based for instance on the reduced variance of the concentration distribution.

A second remark is that an experimentally measured degree of mixing depends on the spatial resolution of the probe used for measuring "point" values of the concentration.

In liquids, the most commonly used methods are electrical conductivity (18, 19), light absorption, fluorescence (30) and chemical methods based on the color change of an indicator under the influence of an instantaneous reaction (21, 22). The spatial resolution of physical methods (optical, electrical microprobes) is about 100 μm (19) so that these are well suited to macromixing studies but cannot compete with chemical methods for the study of mixing at the molecular scale. An original method based on the continuous injection of radioactive tracers in an industrial mixer has also been proposed (23). In gases, concentration fluctuations have been measured using a catalytic wire (24).

The definitions of the degree of mixing presented above aim at a local characterization of the mixture homogeneity in the physical space. There also exist more indirect mixing indices. The segregation index J of Danckwerts (12) is one of the most famous ones. It applies to continuous reactors and relies upon the variance of age : $J = \text{Var}\ \alpha_p/\text{Var}\ \alpha$, where α is the age of a molecule,

α_P the mean age at "point" P and Var denotes the variance of the distributions over all "points" or all molecules in the reactor. J = 1 in a totally segregated mixture (all the ages are the same within a "point" $\alpha = \alpha_P$) and J takes a minimum value depending on the residence time distribution when mixing earliness is at a minimum. J is still a very popular quantity in the literature where it is used by many authors to compare their micromixing models. However I think that whilst it constitutes some useful reference for mixing earliness, it is of little interest for the design of chemical reactors as it cannot be measured directly nor be used for a straightforward calculation of chemical conversion.

Another indirect mixing index was proposed by Ogawa et al. (25, 26), based on information theory and on the estimation of the "entropy" of a mixture. This "entropy" is defined from the tracer distribution among n zones in the reactor as $H = -\sum_{i=1}^{n} V_i^* p_i \ln p_i$, where V_i^* is the reduced volume of zone i and p_i the "probability" of occurence of the tracer in zone i (see reference (25) for more details). As mixing proceeds, the degree of homogeneity of the mixture is defined as $M = H/H_\infty$, where H_∞ is the final entropy of the homogeneous mixture. This method was used to study the mixing rate in stirred vessels. M was plotted as a function of time from a record of the concentrations at various places in the vessel. This allowed a comparison of efficiency of different agitators (26). As an interesting generalization, the macromixing homogeneity in a continuous reactor having a RTD E(t) may be defined as (25) : $M = -\int_0^\infty E(t) \ln E(t)\, dt$, from which it is seen that M = 0 in a plug flow reactor and M = 1 in a continuous stirred reactor.

Eulerian approach: Mixing and Turbulence

Turbulence theory provides a classical approach to mixing phenomena. This is a natural way for mechanical engineers and specialists of combustion, who are very familiar with the methods of fluid mechanics. However, when complex chemical reactions are involved, the use of the formalism of turbulence alone seems to lead to a deadlock, as has been pointed out by several authors. An excellent presentation of the state of the art can be found in the recent literature especially by Brodkey (16, 2, 27) and Patterson (3). These reviews reveal no major breakthrough, and only slow progress on a difficult road.

As the results of turbulence theory are sometimes used incorrectly in the literature, it seems helpful to recall (Table I) the principal quantities characterizing velocity and concentration fluctuations, based on the assumption of homogeneous isotropic turbulence. Most of them can be deduced from spectral measurements. If the techniques for determining velocity fluctuation spectra are well established (hot wire anemometer, laser-doppler anemometer...) reliable methods for monitoring concentration fluctuations are less known. Interesting data were recently obtained by using local conductometry microprobes interfaced with a high gain, fast response conductimeter (28, 29, 30), but there are still problems of space and time resolution (19) and new methods should be developed for the determination of concentration fluctuations spectra in chemical reactors. It is clear from Table I that there are two parallel families of characteristics pertaining to velocity fluctuations on one hand and to concentration fluctuations on the other hand. This distinction is not always made in the literature, some authors using for instance velocity macro or microscales, (which are better known) in place of concentration scales. It must also be recalled that different length and time scales have very precise meanings and should not be used at random on the single basis of dimensional analysis. The macroscales L_f and L_s characterize large initial eddies, whose size is generally comparable to that of the impeller in stirred tanks, or that of the inlet tubes in tubular reactors. Taylor and Corrsin microscales λ_f and λ_s are those of the maximum of dissipation either of the turbulent kinetic energy, or of the segregation. In this respect, the Corrsin microscale λ_s and the associated time constant τ_s play an important role. The classical treatment of Corrsin, widely used in the literature on mixing, yields the following expressions :

$$\tau_s = \left(\frac{5}{\pi}\right)^{2/3} \frac{2}{3 - Sc^2} \left(\frac{L_s^2}{\varepsilon}\right)^{1/3} \quad \text{for gases} \qquad (3\text{-}1)$$

$$\tau_s = \frac{1}{2}\left[3\left(\frac{5}{\pi}\right)^{2/3}\left(\frac{L_s^2}{\varepsilon}\right)^{1/3} + \left(\frac{\nu}{\varepsilon}\right)^{1/2} \ln Sc\right] \quad \text{for liquids} \qquad (3\text{-}2)$$

In liquids, the second term of (3-2) is often negligible and $\tau_s \approx 2(L_s^2/\varepsilon)^{1/3}$.

The problem is to estimate the turbulent energy dissipation per unit mass ε from macroscopic data. For instance, in stirred tanks, it is not obvious that all the power P dissipated at the shaft contributes to ε and several authors (2, 31, 32) have been led to introduce an unknown "efficiency" η such that $\varepsilon = \eta P/\rho V$. An additional difficulty is that ε is not uniform within the tank volume and may vary by a factor 10 from one place to the other. (See Section on stirred tanks).

In order to estimate τ_s, Patterson (3) proposed assimilating L_s to L_f and found $(L_s^2/\varepsilon)^{1/3} = (L_f^2/\varepsilon)^{1/3} = q/\varepsilon$.

A typical value for λ_f in an industrial stirred tank is a few millimeters (19).

The smallest size for turbulent eddies is given by the Kolmogorov microscale λ_K. Energy loss below this size only occurs via viscous dissipation. Here also, several scales have been introduced in the framework of turbulence theory, depending whether velocity or concentration fluctuations are considered, namely λ_K, λ_B, and λ_C (see Table I). In liquids, λ_K is typically between 10 and 100 μm. The Kolmogorov microscale λ_K is frequently used in the in-

Table I. Homogeneous Isotropic Turbulence

Velocity $U = \overline{U} + u$		Concentration $C = \overline{C} + c$	
Mean square fluctuation	$u' = (\overline{u^2})^{1/2}$		$c' = (\overline{c^2})^{1/2}$
autocorrelation	$f(r) = \dfrac{\overline{u(x)\ u(x+r)}}{u'^2}$		$g(r) = \dfrac{\overline{c(x)\ c(x+r)}}{c'^2}$
Macroscale	$L_f = \int_o^\infty f(r)\, dr$		$L_s = \int_o^\infty g(r)\, dr$
Kinetic energy of turbulent motion	$q = \frac{3}{2} u'^2$	Segregation intensity	$c'^2/c_o'^2$
Turbulent energy dissipation	$\varepsilon = -\dfrac{dq}{dt} = -\dfrac{3}{2}\dfrac{du'^2}{dt}$	Segregation dissipation	$\varepsilon_s = -\dfrac{dc'^2}{dt}$
Spectra	All directions $E(k)$, One direction $E(k_1)$		$E_s(k)$
	$u'^2 = 10 \int_o^\infty E(k_1)\, dk_1$		$c'^2 = \int_o^\infty E(k)\, dk$
	$\varepsilon = 30\ \nu \int_o^\infty k_1^2\, E(k_1)\, dk_1$		$\varepsilon_s = 2\,\mathcal{D} \int_o^\infty k^2\, E_s(k)\, dk$
Taylor microscale	$\lambda_f^2 = \dfrac{30\,\nu\, u'^2}{\varepsilon} = \dfrac{10 \int_o^\infty E(k_1) dk_1}{\int_o^\infty k_1^2\, E(k_1) dk_1}$	Corrsin microscale	$\lambda_s^2 = \dfrac{12\mathcal{D} c'^2}{\varepsilon_s} = \dfrac{6 \int_o^\infty E_s(k)\, dk}{\int_o^\infty k^2\, E_s(k)\, dk}$
Taylor time constant	$\tau_f = \left(-\dfrac{1}{u'^2}\dfrac{du'^2}{dt}\right)^{-1} = \dfrac{\lambda_f^2}{10\nu}$	Corrsin time constant	$\tau_s = \left(-\dfrac{1}{c'^2}\dfrac{dc'^2}{dt}\right)^{-1} = \dfrac{\lambda_s^2}{12\mathcal{D}}$
	$L_f = \dfrac{\pi E(o)}{u'^2} \sim \dfrac{q^{3/2}}{\varepsilon}$	Small Sc :	$\lambda_s^2/\lambda_f^2 = \mathcal{D}/\nu\ ,\ \tau_s = \tau_f$ (Corrsin)
Viscous dissipation			
Kolmogorov microscale	$\lambda_K = (\nu^3/\varepsilon)^{1/4}$	Corrsin microscale	$\lambda_c = (\mathcal{D}^3/\varepsilon)^{1/4}$
Batchelor microscale	$\lambda_B = (\mathcal{D}\nu^2/\varepsilon)^{1/4}$		
Viscous dissipation time constant	$\tau_K = (\nu/\varepsilon)^{1/2}$		$\tau_c = (\mathcal{D}/\varepsilon)^{1/2}$

terpretation of micromixing phenomena, as will be discussed in a later section.

The general equations for chemical reaction in a turbulent medium are easy to write, if not to solve (2). In addition to those for velocities ($U = \overline{U} + u$) and concentrations ($C_j = \overline{C_j} + c_j$), balance equations for $q = \frac{3}{2}\overline{u^2}$, the segregation $\overline{C^2}$, and the dissipations ε and ε_s can be written (3). Whatever the shape of the reactor under consideration (usually a tube or a stirred tank), the solution of these equations poses difficult problems of closure, as $\overline{\vec{u}\,\vec{u}}$, $\overline{\vec{u}\,c_j}$, $\overline{c_j^2}$, and also $\overline{c_i c_j}$, $\overline{c_i^2 c_j}$ ... in the reaction terms have to be evaluated. The situation is even more complicated when the temperature and the density of the reacting mixture are also fluctuating. Partial solutions to this problem have been proposed. In the case of instantaneous reactions ($t_R << \tau_s$) the "equilibrium assumption" applies : the mixed reactants are immediately converted and the apparent rate of reaction is simply that of the decrease of segregation, with Corrsin's time constant τ_s. For instance, with a stoichiometric proportion of reactants, the extent of reaction X is given by $1 - \sqrt{I_s}$ (2), a simple result which can also be found by application of the IEM model (see (33)).

Another closure approximation relies on the use of higher order moments, but this is almost a deadlock as for only two reacting components, a 13-equation model is required ! (3).

A more fruitful solution to the closure problem is provided by the use of probability density functions for the fluctuating components. Various shapes (spiked, square wave, gaussian distributions) have been successfully tried (3).

Among experimental studies of chemical reactions in turbulent media, fast reactions in tubular reactors with multijet injection of reactants are very popular, since the first experiments of Mao and Toor (34) and Vassiliatos and Toor (35). Their data have been (and are still) extensively exploited for testing theoretical models, although one may ask if homogeneous isotropic turbulence was perfectly controlled in these experiments. In order to rule out this objection, a new series of experiments was recently performed by Bennani et al. (28, 29, 30, 36) in a 0.29 m i.d. tube eliminating the influence of boundary layers. Turbulence was created by a grid and carefully controlled by velocity fluctuation measurements. Previous studies (2) had confirmed that the decrease of I_s with a non-reacting species (passive scalar) obeys Corrsin's equation :

$$\frac{dI_s}{dt} = - \frac{I_s}{\tau_s} \tag{3-3}$$

where τ_s is given by (3-2). In this case, Bennani et al. (28) found $\sqrt{I_s} \sim (z - z_o)^{-0.60}$ and a spectrum decreasing as $k_1^{-5/3}$.

Let us now consider the case of a moderately fast second order reaction A + B → products, $r = k_2\, C_A C_B$. The chemical production term involves the correlation $\overline{c_A c_B}$:

$$\mathcal{R}_A = \mathcal{R}_B = -\bar{r} = -k_2\,(\overline{C_A}\cdot\overline{C_B} + \overline{c_A c_B}) \qquad (3\text{-}4)$$

It is found experimentally that $\overline{c_A c_B} < 0$. Mc Kelvey et al. (37) have shown that $\overline{c_A c_B}$ can be calculated when the segregation I_s of a non-reacting species was known, in setting

$$I_s = -\frac{\overline{c_A c_B}}{\overline{C_{Ao}}\cdot\overline{C_{Bo}}} \qquad (3\text{-}5)$$

This is very encouraging because it indicates that Corrsin's time constant τ_s is a basic quantity allowing the prediction of the segregation both in the absence and in the presence of chemical reactions. Bennani et al. have brought additional information in studying concentration fluctuations in their tube, in the presence of a chemical reaction : the alkaline hydrolysis of methyl formate (29). The sodium hydroxyde concentration C_A was monitored by fast conductimetry. $\overline{c_A^2}$ was found to first increase then to decrease as $z^{-0.6}$ while its spectrum exhibited an unexpected decay in k_1^{-3}. Using a closure with a gaussian probability density function, they showed (30) that physical mixing ($1/\tau_s$) and chemistry act in an additive manner in the decay of segregation. This supports lumped models of the IEM type (see Section on phenomenological models below), where these two processes are uncoupled. In addition, Bennani et al. (36) were able to predict the value of the correlation $\overline{c_A c_B} < 0$ from their experiments with a passive tracer. This confirms Toor's hypothesis according to which the influence of turbulence on a chemical reaction with unmixed reactants can be determined from the sole knowledge of the segregation decay of a non reacting species in the same conditions.

We have seen that the formalism of turbulence theory (when used alone) led to practically unmanageable equations. Some authors have tried to avoid the closure problem by introducing phenomenological parameters. As an example, Berker et al. (38) attempted to model $\overline{u c_A}$ and $\overline{u c_B}$ in a tubular reactor with a transversal dispersion coefficient. A more general conclusion of the above results is that Corrsin's theory and Toor's hypothesis provide the framework for a simple model where turbulence causes mass transfer with a time scale $\tau_s \sim (L_s^2/\varepsilon)^{1/3}$, whereas chemical reactions would simply be superimposed to this mass transfer. This led several authors to simulate segregation decrease via a coalescence-dispersion (C-D) process, whose frequency ω only depends on the local value of L_s and ε. Thus Canon et al. (39) successfully interpreted Vassiliatos and Toor's experiments by a C-D model where the fluid was divided into small cells undergoing periodic and random coalescence-redispersions, with the frequency ω, the reaction going on between each coalescence. Canon (39) and Patterson (3) fitted the experimental data with the expression $\omega = 1333(\varepsilon/L_s^2)^{1/3}$ (the numerical coefficient should be used with some precaution, as the result is not exactly expressed in this way in the original paper). This is the basis for a family of C-D models, which will be discussed further. In particular, mixing in a stirred tank can be represented by a set of small cells connected by specified flow-rates, in which mixing occurs via a C-D process whose frequency is given by the expression above (HDM model (40)).

According to these models, the turbulent field for chemical reaction is determined once L_s and ε are known at any position in the reactor. This leads to interesting scale-up criteria in stirred reactors (2) : the mixing characteristics are the same if ε/L_s^2 is kept constant. As $\varepsilon \sim P/\rho V$ and $V \sim L^3$, the power input P should be proportional to $L^3L_s^2$. This may be as high as L^5 if one assumes that $L_s \sim L$. The empirical exponent actually lies in between 3 and 5.

Thus, after starting from the ambitious and general formulation provided by the Eulerian equations of turbulence theory, most authors have been led to introduce simplifying assumptions and closure hypotheses which gradually brought them nearer to the Lagrangian approach and its phenomenological coefficients, as will be discussed in the next Sections. This does not mean that turbulence theory is useless in this matter : it provides the physical interpretation of the phenomenological coefficients and may yield their value as a function of experimental parameters. In addition, the consideration of the sign of concentration correlations yields useful rules to determine the effect of segregation on selectivity (41).

In this Section devoted to turbulence and chemical reaction, recent work on what has been called "chemical turbulence" must also be mentioned. This work is essentially experimental and has been carried out with the famous "Belousov-Zhabotinsky" reaction. Decreasing the space time in a CSTR, some authors (42) have observed a transition from periodic to chaotic concentration fluctuations. They claim that this is a purely chemical phenomenon which has nothing to do with any coupling between diffusion and reaction. This conclusion should be checked by designing experiments where mixing conditions and physical uniformity of the mixture are to be carefully controlled. In any case, the Belousov Zhabotinsky reaction, once its kinetic mechanism is entirely elucidated and well accepted, should be an interesting tool for the study of micromixing in laboratory reactors, owing to its sensitivity to concentration gradients (43).

In conclusion to this Section devoted to the turbulence approach to mixing the main points are :

- Need for more data on spectra E_s of concentration fluctuations.

-Improvement of the techniques for monitoring local instantaneous concentrations down to the viscous dissipation microscale (e.g. spatial and time resolution of conductivity probes), development of new techniques (e.g. optical, radioactive tracer methods).

- Clear distinction between macromixing effects (large scale motions due to average velocities) and micromixing effects (turbulent fluctuations).

- Check for homogeneous isotropic turbulence in test experiments (influence of boundary layers, perturbations close to impellers).

- Need for more theoretical or experimental information on the "efficiency η" of utilization of macroscopic dissipated power P at the molecular scale (ε).

- Corrsin's theory and its characteristics time constant $\tau_s \sim L_s^2/\varepsilon$ constitutes a sound basis for representing segregation in turbulent reacting media. This supports the Lagrangian approach to mixing phenomena (e.g. coalescence, dispersion with $\omega \sim 1/\tau_s$).

Turbulence and combustion

This domain is traditionally at the frontier of Chemical Reaction Engineering, and although being the subject of active research won't be reviewed in detail here. However, it represents an important field of application for the theory of chemical reactions in turbulent media with the additional difficulty that the kinetic schemes involved are often complex (e.g. free radical reactions) and that temperature and density fluctuations have to be taken into account. A major problem is the prediction of the formation of pollutants like CO, NO, NO_x (44-47). Classical treatments using turbulence theory are available (48, 49, 50). It is interesting to notice that authors faced with the closure problem have independently imagined models of the Lagrangian type where interaction between turbulent eddies is accounted for by an apparent mass transfer between each eddy and its average environment (See Spalding and his EBU model (50). This is exactly the same approach as in the IEM model developed in Chemical engineering. Here also, most authors recognize that the rigorous closure problem comes to a deadlock (46) and plead for simple models (51) allowing easier predictions (52). It is very encouraging to see that the chemical engineering approach relying on the RTD concept and describing microscopic mixing via C-D processes accounts well for the production of pollutants, ignition phenomena and stability of combustors (53, 54). This should be an incentive for chemical reaction engineers to develop their original methods and to apply them in neighbouring fields, in cooperation with the specialists of these disciplines who may enrich them with their own experience.

Macromixing and Residence Time Distributions (RTD)

The RTD concept is now well established and widely used for designing, scaling-up and optimizing chemical reactors. This subject was recently reviewed by Nauman (1). Most classical textbooks (55) mainly deal with the simple case of steady state incompressible flow, single inlet and outlet, and purely convective motion of the fluid across the inlet/outlet sections. A useful representation of this situation is the "Bundle of Parallel Tubes" (BPT) model (figure 1) where the reactor volume is reorganised in the form of a bundle of small tubes of increasing length, equal to the residence time t_s. The fluid flows at constant velocity (the particles of the same age are on the same vertical line) and the elementary flowrate dQ in each tube is such that $dQ = E(t_s)\, dt_s$. The locus of the tubes extremity then pictures the familiar $F(t_s)$ curve. This model is introduced because it has interesting properties (56) and provides a concrete picture of the RTD so that a frequent use of this representation will be made in the next Sections.

Recent work has attempted to generalize RTD theory to less restrictive assumptions, so that it might apply to more realistic situations encountered in practice. The type of flow and the operating conditions of the reactor are given in Table II below, together with the corresponding references :

Table II : Generalization of the RTD theory

	Classical assumption	Generalization	Reference
Flow regime	steady	transient, periodical	(1)(56)
Volume	constant	variable, periodically variable	(1)(56) (58)(59)
Density	constant	variable	(1)(56)
Inlet/Outlet	convective	dispersive	(1)(57)
Number of Inlets/Outlets	1/1	2/1 n/n	(1)(56)(60) (61)(62)(1)
Semi-batch reactor		1/0 2/0 }	(56)

RTD problems in variable regimes can be dealt with as special cases of the population balance method (56).

Let $\psi d\alpha dV$ be the number of particles in volume dV within the age interval $\alpha, \alpha + d\alpha$. The total number of particles in the reactor is $N = \int_{\alpha,V} \psi d\alpha dV$ and ψ obeys the well known population balance equation

$$\frac{\partial\psi}{\partial t} + \nabla.(\psi\vec{U}) + \frac{\partial\psi}{\partial\alpha} = 0 \qquad (5\text{-}1)$$

The local age distribution function is defined as $f = \psi V/N$. f is

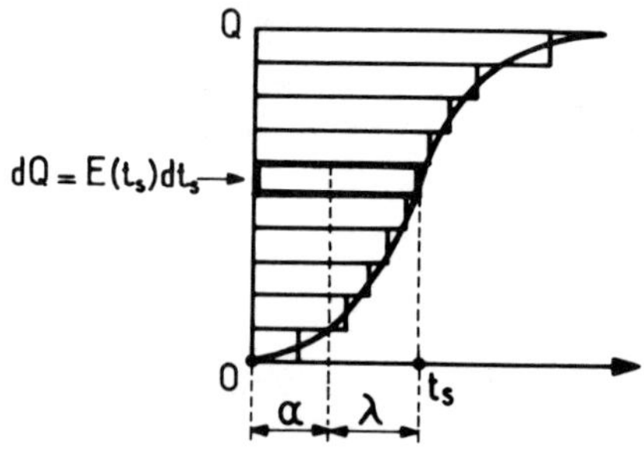

Figure 1. The BPT (bundle of parallel tubes) model. The flow-rate in a tube with residence time between t_s and $t_s + dt_s$ is $dQ = E(t_s)dt_s$. The fluid flows in the tubes at constant velocity. All particles on a vertical have the same age. The height of the pile is Q. The locus of tube extremities is a picture of F-curve $(F = \int_o^{ts} E(t)\, dt)$. Olson and Stout also make use of such a model in their review (177).

normalized and could be the impulse response to an injection of tracer at the inlet. Let $\bar{f}$ be the average value of f over the reactor volume and $\bar{f}_{in}$ and $\bar{f}_{out}$ be the flux averages over the inlet and outlet sections respectively. Then $\bar{f} = I(\alpha,t)$ is the instantaneous internal age distribution (IAD), $\bar{f}_{in} = \delta(\alpha)$ and $\bar{f}_{out} = E(t_s,t)$ is the instantaneous RTD. From (5-1), it is easy to show by averaging that

$$\frac{1}{N}\frac{\partial}{\partial t}(NI) + \frac{Q_{out}\,E - Q_{in}\,\delta(\alpha)}{V} + \frac{\partial I}{\partial \alpha} = 0 \qquad (5\text{-}2)$$

The corresponding moment equation is

$$\frac{1}{N}\frac{d}{dt}(N\,\overline{\mu_n}) + \frac{Q_{out}\,\overline{\mu_{n,out}}}{V} - n\,\overline{\mu}_{n-1} = 0 \qquad (5\text{-}3)$$

where $\overline{\mu}_{n\ out}$ is the nth order moment of $E(t_s,t)$ and $\overline{\mu_n}$ that of $I(\alpha,t)$.

These equations allow various special cases to be treated. This links the nth order moment of the RTD to the $n-1^{th}$ order moment of the IAD.

The IAD $I(\alpha,t)$ in a chemical reactor is specially interesting and it does not seem that sufficient attention has been paid to the possibilities offered by this function. For instance, let us consider a semi-batch reactor, and let Q(t) be the feed flowrate of an incompressible fluid. The instantaneous fluid volume is $V = \int_o^t Q(t')dt'$, from which the IAD is written $I(\alpha,t) = Q(t-\alpha)/V$. $I(\alpha,t)$ can be used to calculate the chemical conversion in different segregation states. Consider a species of concentration C produced with the rate $\mathfrak{R}$. If the mixture is assumed to be well mixed at the molecular scale, one obtains the familiar mass balance equation

$$QC_{in} + \mathfrak{R}V = \frac{d}{dt}(VC) \qquad (5\text{-}4)$$

whereas if the fluid is totally segregated, C varies in each particle as $C_B(\alpha,t)$ (batch conversion) such that :

$$\frac{dC_B}{d\alpha} = \mathfrak{R}, \quad \alpha = 0, \quad C_B = C_{in}(t - \alpha) \qquad (5\text{-}5)$$

The average concentration in the reactor is then given by

$$C(t) = \int_o^t C_B(\alpha,t)\; I(\alpha,t)\; d\alpha \qquad (5\text{-}6)$$

The conversion in partially segregated fluid can be calculated in a similar manner, for instance by using the IEM model (see next Section and (56)). The IAD approach can also be generalized to the case of several feedstreams (56).

Nauman (1) has proposed an interesting treatment of RTD in unsteady-state reactors. Two washout functions are defined :

$W_{in}(t_s,t) = 1 - F_{in}$, fraction of material entering at time t which will remain in the reactor for a duration greater than t_s, and $W_{out}(t_s,t) = 1 - F_{out}$, fraction of material leaving at time t which remained in the reactor for a duration greater than t_s. From these functions, two RTD can be defined, namely $E_{in} = \partial F_{in}/\partial t_s$ and $E_{out} = \partial F_{out}/\partial t_s$ which have all the classical properties of steady state RTD except that they vary with time. In particular, chemical conversion can be calculated in the two limits of mixing earliness (see next Section). For minimal mixedness :

$$C_{out}(t) = \int_0^\infty C_B\left[t_s, C_{in}(t - t_s)\right] E_{out}(t_s,t)\, dt_s \qquad (5\text{-}7)$$

and for maximal mixedness :

$$\frac{\partial C}{\partial \lambda} = -\mathcal{R} + \frac{\left[C - C_{in}(t-\lambda)\right] E_{out}(\lambda,t)}{W_{out}(\lambda,t)} \qquad (5\text{-}8)$$

$$C_{out}(t) = C(\lambda = 0, t)$$

which is a generalized form of Zwietering's equation.

Thanks to this formalism, periodic regimes can be dealt with (estimation of average conversion). The case of an unsteady CSTR has been especially studied (1). Mixing effects in variable-volume chemical reactors were considered by Roth et al. (58). In their model, the volume of a semi-batch reactor is varied periodically (at constant flowrate Q) between V_o and V_m. The filling and emptying stages are separated by a purely batch period. The model depends on two parameters, namely $v = V_o/V_m$ and $\sigma_B = t_{batch}/t_{total}$. The authors show that this periodic operation is equivalent to that in a continuous reactor whose degree of macromixing can be varied between that of a plug flow reactor and that of a CSTR.

In steady state continuous reactors, an important case encountered in practice is that of mixtures with variable density. The general population balance equations presented above theoretically give the solution to the problem if the RTD approach is retained for calculating chemical conversion. However, there exist simpler treatments applicable to special cases. Nauman (1) proposed a solution valid for segregated mixtures whose density only depends on an axial coordinate in the reactor. He used a variant of the Bundle of Parallel Tubes model in which all the tubes have the same length but different flowing velocities. Villermaux (56) extended the well-known tanks in series model to the case where the flowrate Q is variable from tank to tank, according to an expansion factor $\gamma_k = Q_k/Q_{in}$. If the expansion is ignored, the exploitation of a RTD experiment leads to an apparent number of tanks in series larger than the actual one. It would be interesting to see whether this apparent number can be used to obtain an approximate value of the conversion.

The case of reactors with two inlets was considered more than ten years ago by Treleaven and Tobgy (60). The BPT model allows an interesting representation of the flow in these reactors (56) (figure 2). This will be useful when micromixing phenomena will be dealt with (next Section).

A somewhat academic problem also seems to have found its solution in the work of Nauman (57) who showed that RTD in systems governed by the dispersion equation could be defined as in closed systems provided the time spent by the tracer outside the reactor owing to diffusion be excluded from the count.

The last item deserving emphasis in this section is the promising concept of "generalized reaction time" (1) (56). Consider a chemical reaction with rate equation $r = k\ F(C_j)$ and assume that k is a locally variable parameter owing to spatial distributions of external porosity ε_e (catalytic reaction), temperature, etc... and let $\bar{k}$ be the average value of k over the whole reactor. Defining the local parameter $h = k/\bar{k}$, it is possible to introduce a generalized reaction time t_i such that $dt_i = h\ dt_s$, where h is the local value of $k/\bar{k}$ along the tube of the BPT model (figure 1). The value of t_i for a particular tube (residence time t_s) is $t_i = \langle h \rangle\ t_s$, where $\langle h \rangle$ is the average value of h along the tube axis. Now considering all the tubes, there exists a "reaction time distribution" $\mathcal{H}(t_i)$ similar to the RTD $E(t_s)$. The conversion along the tube can be calculated according to

$$C_o \frac{dX}{dt_i} = k\ F(C_j) \tag{5-9}$$

and the overall extent of reaction (under the assumption of minimal mixedness) is

$$X = \int_0^\infty X(t_i)\,\mathcal{H}(t_i)\ dt_i \tag{5-10}$$

This generalizes the well-known equation established in the case of RTDs. Notice that $t_i = t_s$.

For homogeneous isothermal reactors, $h \equiv 1$ and $\mathcal{H}(t_i)$ is the RTD. For isothermal heterogeneous reactors, $h = (1 - \varepsilon_e)/\varepsilon_e$ and $\mathcal{H}(t_i)$ is the contact-time distribution (CTD) introduced by Nauman and Collinge (1). In the case of non-isothermal reactors, $h = k(T)/\overline{k(T)}$, and $\mathcal{H}(t_i)$ is a "thermal time distribution".

Unfortunately, if there exist experimental methods for measuring the RTD and the CTD by means of tracer tests, nothing has been published about the experimental determination of the latter distribution. At present, this limits the potential applications of such an approach.

This brief review on RTDs would be incomplete without a few

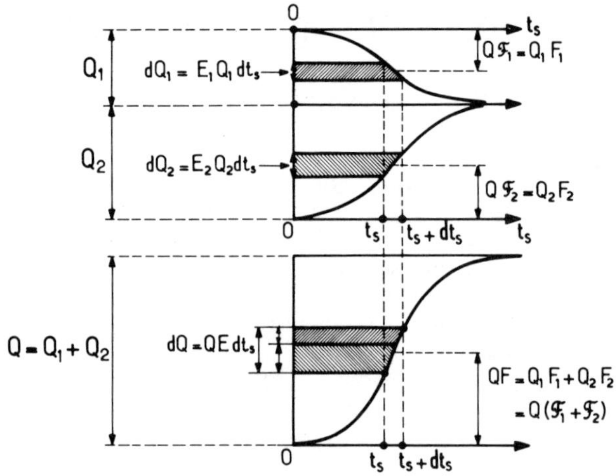

Figure 2. Extension of the BPT model to the case of a reactor having two inlets (two unmixed feedstreams of respective flowrates Q_1 and Q_2). The picture shows how to build the overall RTD from those of each separate feedstream by associating tubes having the same residence time.

words on experimental techniques. Tracer methods are now routinely utilized in industry and in laboratories. Bourne et al. (63) described a reversible photochromic tracer for RTD measurements in highly viscous liquids. Conochie and Gray (64) developed a spectrophotometer for the study of mixing in a channel simulating a metallurgical reactor. Interesting visualisation techniques were also devised by Quraishi and Fahidy (65) who induced the change of color of an indicator at a surface by electrolysis and by Castellana et al. (66) who studied mixing in reactor vessels by means of a gamma camera and a Technetium 99 tracer. There is still a need for a good tracer for particulate solids. Magnetic tracers should be developed for this purpose (9).

In conclusion to this section, research in the RTD area is always active and the initial concepts of Danckwerts are gradually being completed and extended. The population balance approach provides a theoretical framework for this generalization. However, in spite of the efforts of several authors, simple procedures, easy to use by practitioners, would still be welcome in the field of unsteady state systems (variable volumes and flow rates), multiple inlet/outlet reactors, variable density mixtures, systems in which the mass-flowrate is not conserved, etc... On the other hand, the promising "generalized reaction time distribution" approach could be developed if suitable experimental methods were available for its determination.

Mixing earliness in the age domain : phenomenological models

The RTD (one or two inlets) characterizes macromixing, but it is well known that this is not sufficient to describe the whole mixing process in the reactor. Additional information is required and two questions must be answered : how are particles of different ages mixed and brought into contact, taking into account the RTD constraint ? This is the problem of "mixing earliness", the subject of this Section. Once particles having the same life expectancy (or residual lifetime) are close to each other, what is the degree of mixing between molecules contained in these particles ? This is the problem of "segregation", discussed in the next Section. It must be clearly specified that these two problems are distinct and independent, a point which does not seem to be obvious in the mind of all authors. In other words a fluid in a state of maximum mixedness may well remain segregated if microscopic interaction between particles is uneffective. In a continuous reactor, the fluid is in a state of Minimum Mixedness (Min Mix) if mixing between particles occurs at the latest moment : Neighbouring particles have the same age. Conversely the fluid is in a state of Maximum Mixedness(Max Mix) if particles of different ages are mixed as soon as they enter the reactor, mixing occurs at the earliest moment and neighbouring particles have the same life expectancy. Physical mixing among particles is characterized by a different notion : the state of segregation which may lie between two extreme situations : the "macrofluid", where no physical mixing at all takes place between particles, and the "microfluid", which is intimately mixed on the molecular scale. There is an ambiguity between these two notions in the literature because many authors make the implicit assumption that once the fluid is mixed in the age domain, it behaves like a microfluid i.e. minimum mixedness is confused with segregation and maximum mixedness with molecular mixing.

In what follows, it may happen that we accept this assumption in order to explain their treatments but we must keep in mind that this is not necessary.

We may now recall the fundamental equations for calculating chemical conversion in the limiting states of Min Mix and Max Mix. The BPT model provides a convenient picture of the two situations (figure 3). If the bundle is arranged in such a way that the particles of same residual lifetime are situated on the same vertical line, then minimal mixedness corresponds to a perfect insulation between tubes and the conversion for a single reaction is written

$$\begin{cases} C_j = \int_0^\infty C_{j,B}(\alpha)\ E(\alpha)\ d\alpha \\ \dfrac{dC_{j,B}}{d\alpha} = \mathcal{R}_j(C_k) \quad ,\alpha = 0 \quad C_{j,B} = C_{jo} \end{cases} \tag{6-1}$$

Conversely, if the particles on a same vertical line are really mixed, then maximum mixedness is achieved. In addition, if the fluid is well micromixed, then, Zwietering's equation applies :

$$-\frac{dC_j}{d\lambda} = \mathcal{R}_j + (C_{jo} - C_j)\ \frac{E(\lambda)}{1 - F(\lambda)} \tag{6-2}$$

to be integrated from $\lambda = \infty$ to $\lambda = 0$. The problem of modeling intermediate mixing states has been and is still the subject of many theoretical investigations. Equations (6-1) and (6-2) are written at steady state. Chen (67) has shown how to generalize them to unsteady regimes by using the population balance formalism. An excellent review of all the models proposed for representing impremixed feedstreams reactors was published by Ritchie and Tobgy (6).

It is useful to try to set some order in this host of models. For this purpose, the representation proposed by Spencer, Leshaw et al. (68) is helpful. It is essentially a two-environment model in which the assumption cited above: Min. Mix. = macrofluid and Max. Mix. = microfluid is implicitly made. Along the axis of a small tube of the BPT model, the fluid gradually passes from a Min. Mix. state to a Max. Mix. state (Figure 4). The residence time in this particular tube lies in the range t_s, $t_s + dt_s$ and the flowrate is $dQ = Q\ E(t_s)\ dt_s$. The flow transferred from the Entering Environment (E.E.) to the Leaving Environment (L.E.) in the interval

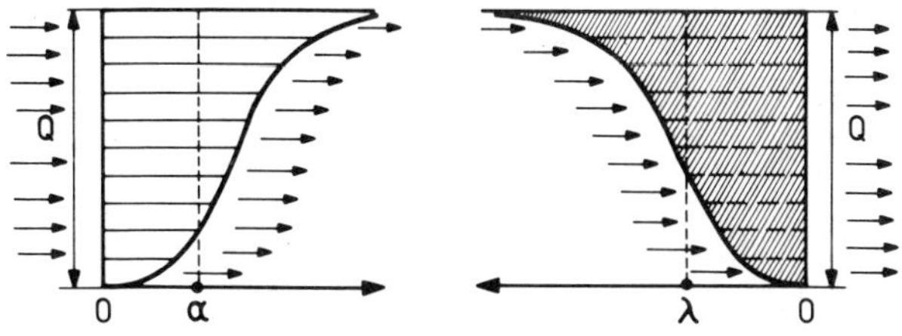

Figure 3. The two limiting cases for mixing earliness. Left, Min. Mix.: particles having the same age are close together (on the same vertical); right, Max. Mix.: particles having the same life expectancy are close together (on the same vertical).

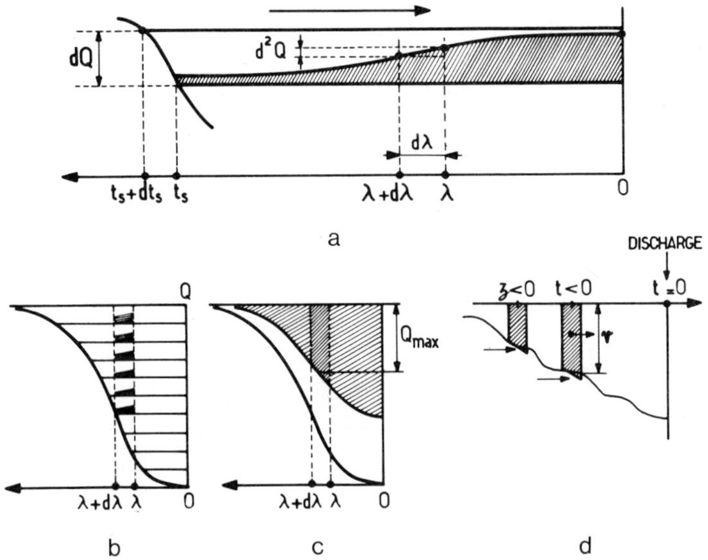

Figure 4. Model of Spencer and Leshaw (68). Along the axis of a tube of the BPT model, the fluid gradually passes from a Min. Mix. E.E. (unshaded) to a Max. Mix. L.E. (shaded). Mixing is characterized by the function $h(\lambda, t_s)$ such that over $d\lambda$, $d^2Q = h(\lambda, t_s)d\lambda dQ$ (a). All the Max. Mix. fluid having the same life expectancy λ can be collected (b) in a single zone with flowrate Q_{Max} (c). This zone may also be considered as an "accumulator" (dimensionless volume v) transiting along the time axis and receiving the Max. Mix. fluid before being discharged at time t = 0 (d).

$d\lambda$ is : $d^2Q = Q\ A(\lambda,t_s)d\lambda dt_s$ where $A(\lambda,t_s)$ is a mixing function which can also be expressed from the RTD by introducing $h(\lambda,t_s)$: $A(\lambda,t_s) = h(\lambda,t_s)\ E(t_s)$. For a given life expectancy λ, the fraction of fluid in the Min.Mix. state is s and the complementary fraction 1 - s is in the Max.Mix. state. From (6-4) and figure 4, $s(\lambda,t_s) = s(0,t_s) + \int_0^\lambda h(\lambda,t_s)d\lambda$ (6-3) where one may have $s(t_s,t_s) \neq 1$ (tube inlet) and $s(0,t_s) \neq 0$ (tube outlet). $s(\lambda,t_s)$, also written $s(\lambda,\alpha)$ is known as the "segregation function" although, as the above remark indicates, "segregation of ages" would be better. In our particular tube, the flowrate of the Max. Mix. fluid is $dQ_{Max} = Q\ g(\lambda)d\lambda$ with $g(\lambda) = \int_\lambda^\infty A(\lambda,t_s)\ dt_s = \int_\lambda^\infty h(\lambda,t_s)\ E(t_s)dt_s$. In the whole reactor, the Max.Mix flowrate is thence $Q_{max} = Q \int_\lambda^\infty g(\lambda')\ d\lambda'$.

This transfer between two environments can also be described as a Lagrangian process where a small band of volume $dV_{Max} = Q_{Max}\ d\lambda$ moves with a constant velocity along the λ axis in the direction of decreasing λ. Spencer et al. (68) consider this volume as an "accumulator" whose dimensionless volume v was defined by

$$\frac{dv}{dt} = \int_{z=-\infty}^{t} A\ (-z,\ -t)\ dz \qquad (6\text{-}4)$$

v is the accumulator volume at time $t < 0$.

The accumulator entered the reactor at time $z < 0$ $(-z < -t)$. It leaves the reactor and discharges at time $t = 0$. This Lagrangian formulation is very interesting as it allows the representation of transient regimes by a succession of accumulators which are gradually filled during their transit in the reactor and are discharged when they leave it. The conversion is then given by the following equation :

$$\frac{dC_j}{dt} = \frac{1}{v} \int_{z=-\infty}^{t} A(-z,-t) \left\{ C_{j,B}\left[C_k(z),\ t-z\right] - C_j \right\} dz + \mathfrak{R}_j \qquad (6\text{-}5)$$

$C_{j,B}$ is the familiar batch expression and the equation has to be integrated from $t = -\infty$ $(vC_j \to 0)$ up to $t = 0$.

Most "micromixing models" published in recent years can be compared on the base of their "segregation" function. As their abstract mathematical character may render the original publications difficult to understand to non-specialists, I have tried to illustrate their principle in a more concrete manner by representing both the two environments in the BPT model and the related segregation functions in the (λ,α) space (Figures 4 and 5). The reader will recognize well-known models of "intermediate micromixing." He will also agree that one can easily imagine an almost endless set of models of this kind by varying the shape of the s-function without really bringing in anything new.

Once we know how the fluid is transferred from the Entering

to the Leaving Environment, there remains the need to specify the mechanism of interaction between particles having the same life expectancy in the L.E. This is the problem of so called segregation decay. A first way to represent this interaction is to assume that particles with the same λ undergo random coalescence-dispersions (C-D) with frequency ω. Chemical reactions, if any, occur within the particles between two coalescences. At each coalescence, the molecules in the particles are mixed up and the composition is thus averaged before dispersion. In a continuous reactor with space time τ, the corresponding parameter is $I = \omega\tau$, representing the average number of coalescences per particle during its stay in the reactor. The random-coalescence process is generally simulated by a Monte-Carlo method. This method was first used by Spielman and Levenspiel (82) and then extensively exploited by Kattan and Adler (83) and Ritchie and Tobgy (77).

In a second kind of model, the interaction is represented via a deterministic process. Each particle is assumed to exchange matter with an average environment made up of all particles having the same life expectancy. For instance, the IEM Model (Interaction by Exchange with the Mean), was initially developed for representing micromixing in stirred reactors (33, 32). However, it can easily be extended to any RTD (84). The basic equation for conversion of a species in the L.E. is written

$$\frac{\partial C_j}{\partial \alpha} = \frac{\overline{C}_j(\lambda) - C_j(\lambda,t_s)}{t_m} + \mathcal{R}_j \qquad (6\text{-}6)$$

where $$\overline{C}_j(\lambda) = \frac{1}{1 - F(\lambda)} \int_{\lambda}^{\infty} C_j(\lambda,t_s)\ E(t_s)\ dt_s \qquad (6\text{-}7)$$

The characteristic parameter is the micromixing time t_m (N.B. : $t_s = \alpha + \lambda$).

In a third family of models, interaction between particles is assumed to occur by molecular diffusion, either between a particle and its average environment (81, 109) or between two specified categories of particles (79). This point will be examined with more detail in the next Section.

The case of reactors with two unmixed feedstreams is especially interesting (6), because more realistic for applications. It has been thoroughly investigated in a series of papers by researchers of Exeter University (UK). The BPT model offers a convenient picture of the reactor where both feed streams have their own RTD and the two bundles are placed side by side in such a manner that particles with the same life expectancy are situated as usual on the same vertical line (figure 6). In a first paper (60), a distinction is made between age mixedness and species mixedness : Maximum age and species mixedness is achieved if particles with the same life expectancy are mixed. A more restrictive case is that where particles are able to mix only if they have both the same

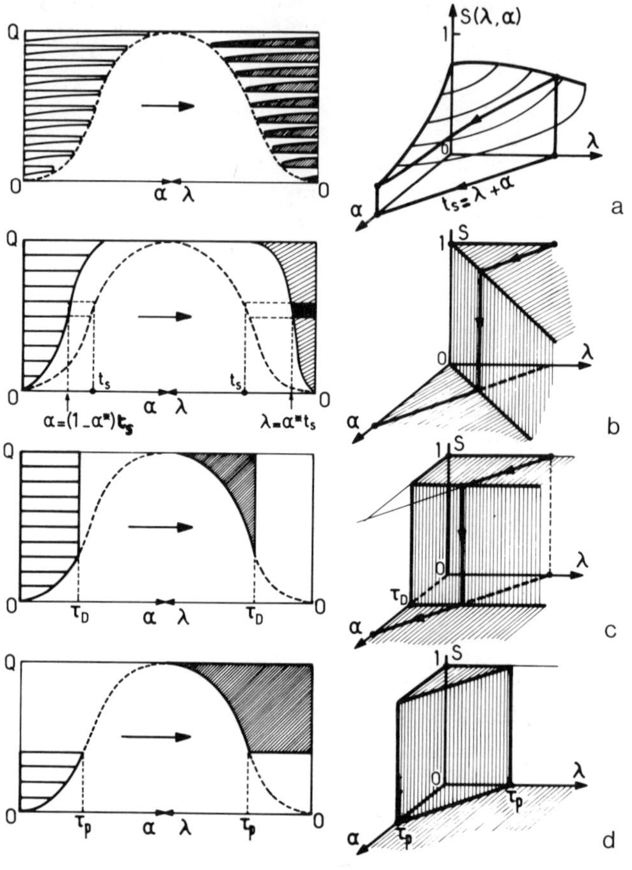

Figure 5. Models. a, general case. Fluid flows from left to right at constant velocity and is transferred from the E.E. (unshaded tubes, Min. Mix.) to the L.E. (shaded area, Max. Mix.). In the $s(\lambda,\alpha)$ space, a particle having a given residence time t_s describes a trajectory between the plane $\alpha = 0$ and the plane $\lambda = 0$; b, $\alpha*$ model of Spencer et al. (68). Each particle stays in the E.E. for a duration proportional to its residence time; c, τ_D model of Spencer et al. (68), which is also the series model of Weinstein and Adler (69). Each particle stays for a constant time τ_D in the E.E., provided $t_s > \tau_D$; d, parallel model of Weinstein and Adler (69). All particles with $t_s < \tau_P$ stay in the E.E. The other ones are immediately transferred in the L.E.

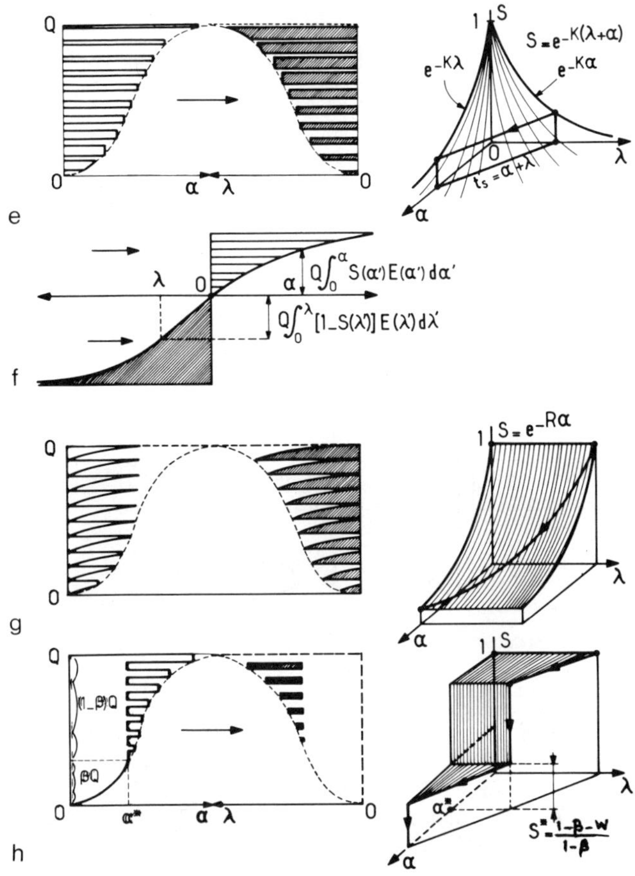

Figure 5. Models. e, model of Villermaux and Zoulalian (70). For each residence time, a given fraction of the particles stay in the E.E. and the rest is immediately transferred in the L.E. The segregation function s only depends on t_s. On the figure, s = exp($-Kt_s$) but other functions may be proposed; f, in Villermaux and Zoulalian's model (70) (when s only depends on t_s), an alternative representation is possible where all the tubes of the E.E. and all those of the L.E. have been piled up together; g, model of Ng and Rippin (72). It is somewhat similar to the preceding one, except that s only depends on the age, s = exp(− Rα). For representing the reduction of size of entering particles in an unmixed feedstream reactor, Plasari et al. (71) have used a model of this kind where s = $(1 - \alpha / t_{\acute{e}})^3$; h, model of Valderrama and Gordon (73 - 76). This is a two-parameter model (β,w) derived from the one on Figure 5c.

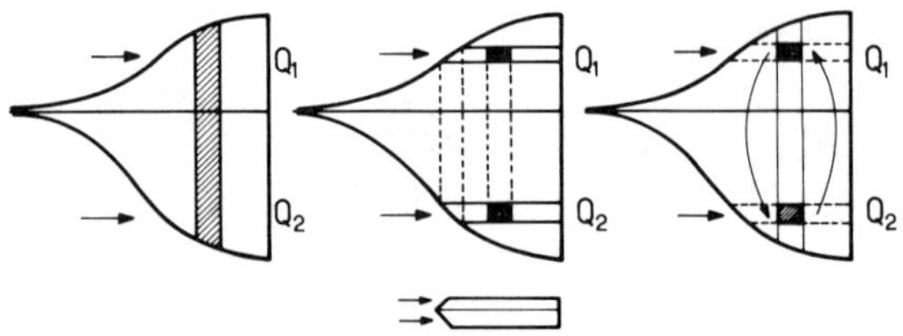

Figure 6. Mixing in a flow reactor with two unmixed feedstreams (60). Left, Maximum age, maximum species mixedness; middle, minimum age, maximum species mixedness (mixing in this case can also be assumed to occur by molecular diffusion between two tubes having the same and (79)); and right, mixing between particles of the same life expectancy by a random-coalescence process (77).

life expectancy λ and the same age α (Minimum age, Maximum species mixedness). Mixing is simulated by Monte-Carlo method (random coalescence) (77). This theoretical work is supported by an experimental study where a reactor was fed with two unmixed feedstreams having different RTD (85). Chemical reaction measurements are reported (86), and it is shown that a single interaction parameter $I = \omega\tau$ accounts for chemical conversion when the stoiechiometric ratio and the Damköhler number ($kC_0\tau$) are varied. The last paper of the series (79) compares the merits of a reaction/diffusion model and of the C-D model. It is concluded that C-D provides a better correlation of experimental results. The models presented on figure 5 can easily be extended to the case of two separate feedstreams by joining the corresponding Entering and Leaving Environments side by side (figure 7). For instance, a three environment model has been proposed by Ritchie and Tobgy (87) (Two E.E. and one L.E.) with the segregation function $s(\alpha) = \exp(-R_s\alpha)$. A generalization to four environments (Two E.E. and two L.E.) was studied by Mehta and Tarbell (88). A Three-Environment model where the segregation function only depends on t_s was published by Ritchie (78). In this case, the representation of figures 5-6 is valid. The model depends on three parameters : two interaction moduli I_E and I_L for the E.E. and the L.E. respectively, and the "segregation" parameter $R_p\tau$ $[s = \exp(-R_p t_s)]$. The reader will understand that with a little imagination an almost endless suite of models of this kind can be conceived.

In the models cited above, interaction between the environments from the two feedstreams was modelled by a C-D mechanism. The IEM Model can also be used for this purpose. Equation (6-6) still holds but the mean value $\bar{C}_j(\lambda)$ at a given life expectancy must comprise both feedstreams 1 and 2 :

$$C_j(\lambda) = \frac{1}{1 - \mathcal{F}_1(\lambda) - \mathcal{F}_2(\lambda)} \left[\int_\lambda^\infty C_{j1}(\lambda,t_s)\, a_1\, E_1(t_s)\, dt_s + \int_\lambda^\infty C_{j2}(\lambda,t_s)\, a_2\, E_2(t_s)\, dt_s \right] \qquad (6\text{-}8)$$

with $a_1 = Q_1/(Q_1 + Q_2)$, $\mathcal{F}_1(\lambda) = a_1 \int_0^\lambda E_1(t_s)dt_s$ and similar definitions for feestream 2.

I now come to a very important observation. We have seen that interaction between the environments (or between the particles themselves if this is assumed equivalent to segregation decay) could be represented by different one parameter models, e.g.

- The C-D model, parameter $I = \omega\tau$
- The IEM model, parameter τ/t_m
- The Models of Ng and Rippin, parameter $R_s\tau$
- The Shrinking-Aggregate Model (71), parameter τ/te ...

Numerical simulations and theoretical arguments (based on concentration variance or chemical conversion) lead to the very inter-

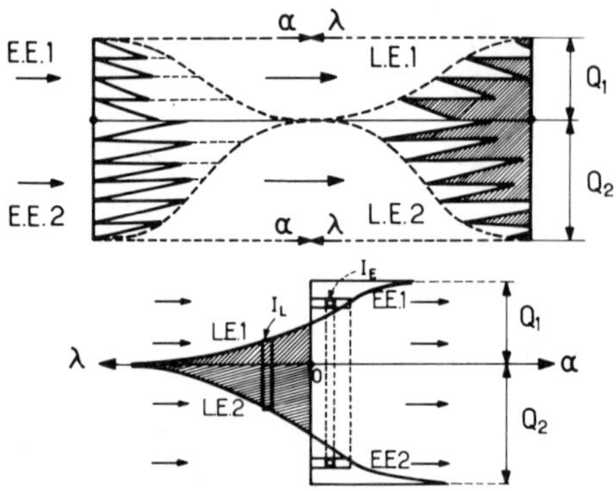

Figure 7. Top, extension of Spencer and Leshaw's model to the case of a reactor having two inlets (unmixed feedstreams). In the general case, there are four environments (two entering, two leaving). Bottom, when the segregation function only depends on residence time, this representation is also valid (78). Three environment model with three parameters R_p $s = \exp(-R_p t_s)$, I_L (interaction modulus in the leaving environment), I_E (interaction modulus in the entering environment, assumption of Figure 6, middle).

esting result that they yield approximately equivalent results if the following relationship between the parameters is established (33, 78) :

$$\frac{\omega\tau}{2} = \frac{I}{2} = \frac{2\tau}{t_m} = R_s\tau = \frac{8\tau}{t_e} \qquad (6\text{-}9)$$

Let us give three examples for this equivalence in comparing the C-D and the IEM models. Figure 8 shows the conversion for an instantaneous reaction in a CSTR fed with unmixed feedstreams in stoichiometric amounts (33). On Figure 9, the yield of intermediate product R in a system of consecutive competing reactions $A+B \rightarrow R$ $R+B \rightarrow S$ is plotted against the conversion of B in a CSTR with unmixed feedstreams (33, 80). Figure 10 compares results obtained for a second-order reaction in a reactor with unmixed feedstreams having different RTDs (equivalent to 2 and 6 tanks in series) by Treleaven and Tobgy (77) to a simulation by the IEM model using equations (6-6) and (6-8)(84). In all the cases, a satisfactory agreement is observed between the predictions of both models. Of course, this would be less good for small values of the interaction parameters (see figure 8), when the particles undergo statistically very few coalescences during their transit in the reactor. We can conclude from this that all these phenomenological models are approximately equivalent, and that the choice in favor of one or the other is often more a matter of computational convenience than of physical relevance. Other models in the literature are based on systems with recycle loops, as for instance in reference (89) where the fluid flows in a plug flow reactor with distributed lateral outlet and recycling at a given age. Dudukovic (90) has shown that these recycle models are not recommended for representing intermediate micromixing as it is difficult to retain the same RTD whilst varying the recycle ratio. However, in using this model and others, he found (91, 92) that the number of multiple steady states could depend on the degree of segregation in chemical reactors where some forms of Langmuir-Hinshelwood reactions were carried out. For each model, some authors still evaluate Danckwert's degree of segregation J and try to link it to the segregation intensity (93, 94) but we have seen above that this is rather an academic exercise.

The last point that will be discussed in this section is concerned with identification of micromixing by using reactive tracers.

It is known that the spread of an inert tracer during its transit in the reactor can only yield information on macromixing (unless concentration fluctuations are measured as explained above). However, owing to its sensitivity to mixing history, the conversion of a reactive tracer may allow identification of intermediate micromixing states. For instance, Parini and Harris (95) proposed to determine the relative position of mixing zones in models having the same RTD by studying the conversion in a pulse

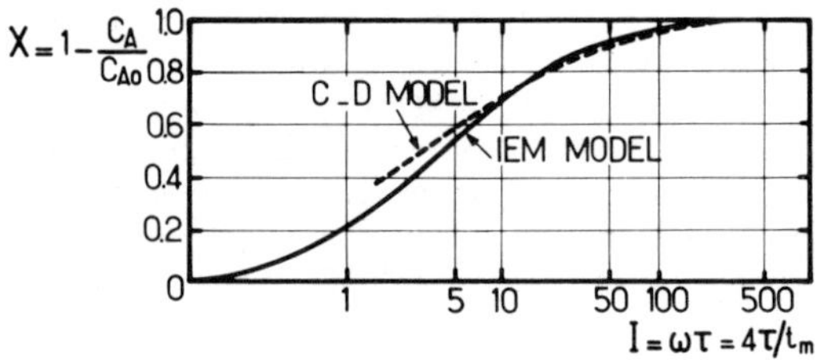

Figure 8. Equivalence between the C-D model and the IEM model. Instantaneous reaction between two unmixed feedstreams of A and B in stoichiometric amount in a CSTR (33).

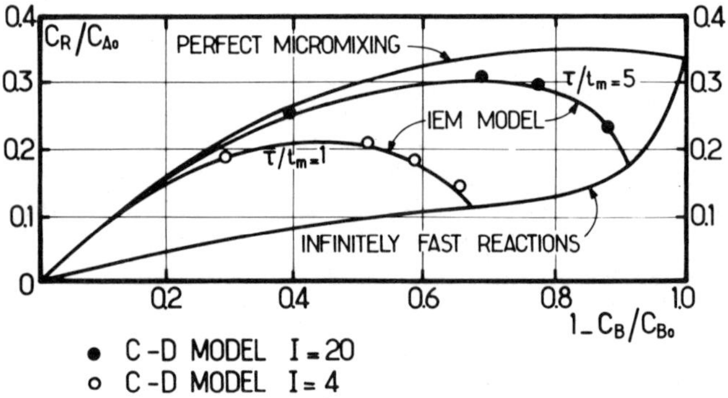

Figure 9. Equivalence between the C-D model and the IEM model. Consecutive competing second-order reactions $A + B \xrightarrow{k_1} R$, $R + B \xrightarrow{k_2} S$. Unmixed feedstreams of A and B in a CSTR. $C_{Ao} = C_{Bo}$, $k_2/k_1 = 0.5$. The yield of R is plotted against the conversion of B (80, 33).

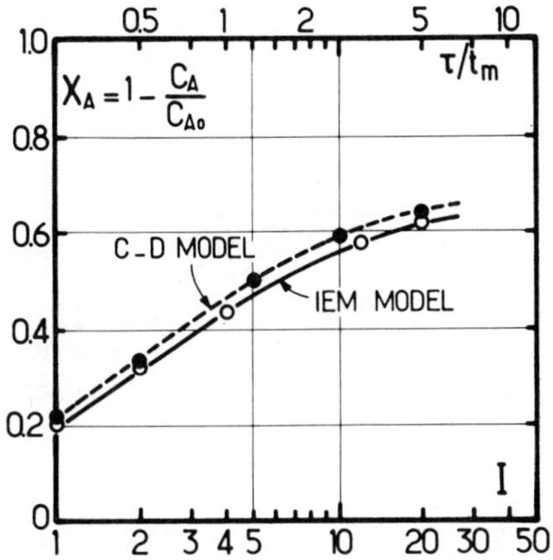

Figure 10. Equivalence between the C-D model and the IEM model. Reactor with two inlets having different RTDs equivalent to that of 2 and 6 tanks in series, respectively. Unmixed feedstreams of A and B (equal flowrates), second-order reaction, $C_{Bo} = 8\ C_{Ao}$, $kC_{Ao}\tau = 0.5$. Data of Treleaven and Tobgy (77) for the C-D model.

of reactive tracer. A similar study was made by Wood and Hop (96). More recently, Spencer et al. proposed and interesting method to determine mixing earliness by using two reactive tracers A and B (68). When the fluid is close to Min.Mix. ($\alpha^* \approx 0$ or τ_D large in the models of figures 5.2 and 5.3), the best discrimination is obtained when A and B are injected in the form of two sharp pulses separated by a short interval. When the fluid is close to Max.Mix. ($\alpha^* \approx 1$, $\tau_D \approx 0$) the recommended injection shape is a square pulse containing both A and B. The conversion in the outlet pulse is then strongly dependent on micromixing. An experimental illustration of the method is given in the next paper by Spencer and Lunt (97). The reactive tracers were p-nitrophenyl acetate and sodium hydroxide which undergo a second-order reaction to produce p-nitrophenylate, an optically detectable product. In a poorly mixed CSTR, segregation was found to increase with viscosity but the interpretation of the experiments is difficult owing to an ill-defined RTD.

The conclusion to this Section is clear : there are now enough phenomenological models in the literature for representing intermediate micromixing (mixing earliness) in flow reactors. Most of them are more or less equivalent. Instead of imagining more and more sophisticated mathematical ways to pass from the Entering to the Leaving Environment, it would be advisable to collect more experimental data on real reactors (the reactive tracer method is especially promising in this way) and to try to correlate the value of the micromixing parameters to physicochemical properties of the mixture and operating characteristics of the reactor. This leads us directly to the next Section.

Segregation and mixing at the microscopic level : mechanism of micromixing in the physical space.

In this section, the fundamental mechanism of interaction between fluid particles will be discussed. Until now, the term "particle" has not been accurately defined. As well a Danckwert's "point", it may denote a fluid blob, clump or aggregate of any shape (sphere, cylinder, slab, lamina) a group of molecules, or more generally any segregated region of the fluid which the mixing process has to destroy and disperse in order to achieve mixing at the molecular scale (or at least at a scale where chemical reactions can occur). This Section will thus be devoted to a Lagrangian approach to the problem already discussed in Eulerian coordinates The aim is to give a physical description amenable to quantitative prediction of segregation decay, relying on experimental parameters.

Simultaneously, such a comprehensive description should provide a physical interpretation for the phenomenological parameters of the models depicted in the preceding Section.

In my opinion, the soundest basis for this description was published by Beek and Miller thirteen years ago (98). According to their own words, three stages may be distinguished in the mixing process :

1. Distribution of one fluid through the other and uniformization of average composition without decreasing local concentration variations.

2. Reduction of size of the regions of uniform composition and increase of contact between regions of different composition.

3. Mixing by molecular diffusion.

Stage 1 is clearly macromixing. Let us review the models proposed for the last two processes (32).

Purely convective mixing. An important contribution to the solution of this problem was given by Ottino, Ranz and coworkers (99-106) in the last three years. However, the abstract character of the mathematical formalism used in these papers might discourage practitioners who are not very familiar with tensor analysis, although the underlying concepts are finally rather simple. The theory considers only convective mixing under the influence of the velocity field. Diffusion is further superimposed on to the pattern created by convection. The main idea is that once the velocity field inside the reactor is known, it is possible to predict the deformation of any material portion of fluid as a function of the Lagrangian time attached to this element. Two quantities are of special interest : the material time rate of change of linear stretch, and that of area stretch. Let $\underline{L} = \nabla.\vec{U} = \frac{\partial U_i}{\partial x_j}$ be the velocity gradient. From this, we may define the stretching tensor $\underline{D} = (1/2)(\underline{L} + \underline{L}^T)$ and the vorticity tensor $\underline{\Omega} = (1/2)(\underline{L} - \underline{L}^T)$. The fundamental relationships established by Ottino, Ranz et al. (99, 101, 103, 105) are then for the rate of change of linear stretch ($\dot{\lambda}/\lambda$) and area stretch ($\dot{\eta}/\eta$), respectively:

$$\dot{\lambda}/\lambda = \underline{D} : \vec{n}\,\vec{n} \quad , \qquad \dot{\eta}/\eta = \nabla.\vec{U} - \underline{D} : \vec{n}\,\vec{n} \tag{7-1}$$

A few words of explanation are not useless in order to understand this formalism. As a consequence of mixing, the medium is assumed to have a lamellar structure and $\vec{n}$ is a unit vector which remains normal to the material slices undergoing deformations in the velocity field. $\vec{n}\,\vec{n}$ denotes a dyadic product (the dyadic product of vectors $\vec{a}$ and $\vec{b}$ is the tensor $a_i b_j$) and $\underline{D} : \vec{n}\,\vec{n}$ denotes the scalar product of the two tensors (the scalar product of tensors $\underline{T} = T_{ij}$ and $\underline{W} = W_{kl}$ is the scalar quantity $\underline{T} : \underline{W} = \sum_i \sum_j T_{ij}\, W_{ji}$). Assume that we start with two miscible fluids A and B (having for instance different colors). Upon mixing, we obtain a lamellar marbled structure characterized by a striation thickness δ and a specific "interfacial" area a_v. If the fluid is incompressible, $a_v\delta = 1$. Then, by application of (7-1)

$$\dot{a}_v/a_v = -\,\dot{\delta}/\delta = -\,\underline{D} : \vec{n}\,\vec{n} \tag{7-2}$$

This equation can be integrated in particular velocity fields to yield the variation of the striation thickness as a function of time (see below). In a newtonian fluid, it can be shown that (99, 105) :

$$\underline{D} : \vec{n}\,\vec{n} \leqslant (\underline{D} : \underline{D})^{1/2} = (\varepsilon_v/2\nu)^{1/2} \tag{7-3}$$

where ε_v is the viscous dissipation per unit mass. If we define a stretching time t_δ as $t_\delta = -\delta/\mathring{\delta}$, (7-3) imposes a lower bound $t_\delta \geqslant (2\nu/\varepsilon)^{1/2}$. (7-3) also allows a local efficiency to be defined as :

$$e = - \frac{\underline{D} : \vec{n}\,\vec{n}}{(\underline{D} : \underline{D})^{1/2}} = - \frac{\underline{D} : \vec{n}\,\vec{n}}{(\varepsilon_v/2\nu)^{1/2}} \tag{7-4}$$

the latter being valid for newtonian fluids only. The average value of e over the whole reactor volume leads to the overall efficiency "eff" of the mixing process telling us how the viscous dissipation is utilized to promote the thinning of the laminae (107, 102, 103, 105). Let us give three examples (the reader may try to find these results by application of (7-2)).

1) Shear flow ($U_x = Gy$, $U_y = 0$, $U_z = 0$, laminae initially normal to Ox) (99, 104) :

$$\mathring{\delta}/\delta = - G^2 t/(1 + G^2 t^2) \text{ and } \delta = \delta_o (1 + G^2 t^2)^{-1/2} \tag{7-5}$$

2) Stagnation flow ($U_x = \alpha x$, $U_y = - \alpha y$, $U_z = 0$, laminae initially normal to Oy) (104)

$$\delta = \delta_o \exp(- \alpha t) \tag{7-6}$$

3) Stretching at constant velocity ($U_x = G_x/(1 + Gt)$, $U_y = 0$, $U_z = 0$)

$$\mathring{\delta}/\delta = - G/(1 + Gt) \text{ and } \delta = \delta_o (1 + Gt)^{-1} \tag{7-7}$$

This kind of stretching occurs for instance in "taffy pull". If the sheet is folded up after a given stretching duration t_o and the resulting sheet is streched up again, then the number of folds is $2^{t/t_o}$ and $\delta = \delta_o\, 2^{-t/t_o}$. In all these processes, efficient mixing is achieved when the laminae are periodically reoriented with respect to the direction of stretching (101). This remark is also important in the design of static mixers.

Erosive (or dispersive) Mixing. Ottino's treatment assumes continuous motion, namely connectedness of material surfaces, and hence conservation of topological features (105). Conversely, one may think of a mixing process that would gradually pull off smaller fragments from the segregated clumps by turbulent friction at their external surface. This is the basis for the "Shrinking Aggregate" (SA) Model of Plasari et al. (71). The peeling-off process is characterized by a mass transfer coefficient h which is assumed to be expressed by the Calderbank-Moo Young correlation, applicable to small particles immersed in turbulent media : $h\ell/\mathcal{D} = 0.13\ \ell(\varepsilon/\nu^3)^{1/4}\ Sc^{1/3}$, ℓ is the diameter of the shrinking aggregate. For spherical shape, this leads to a linear size decrease as a function of age : $\ell = \ell_o(1 - \alpha/t_e)$ with the characteristic erosion time-constant

$$t_e = \frac{\ell_o\ \lambda_K}{0.26\mathcal{D}}\ Sc^{-1/3} \approx \frac{\ell_o\ \lambda_K}{2.6\mathcal{D}} \qquad (7\text{-}8)$$

in liquids (Sc $\approx$ 1000). λ_K is the familiar Kolmogorov microscale (see above). The dependence of t_e on experimental parameters is thus $t_e \sim \ell_o\ \nu^{5/12}\ \mathcal{D}^{-2/3}\ \varepsilon^{-1/4}$. Two problems remain to be solved : How to estimate the initial size ℓ_o of the aggregates ? Is it proportional to the initial velocity U_o of the fluid as it has been assumed in stirred reactors (71) ? What is the expression of the energy dissipation rate ? As already pointed out, it seems necessary to introduce efficiencies η_1 and η_2 for the utilization of mechanical energy and kinetic energy dissipation respectively (71, 108) $\varepsilon = \eta_1\varepsilon_1 + \eta_2\varepsilon_2$. These are questions for further research.

Coalescence-Dispersion Mixing. This process has already been mentioned in the preceding Sections. The particles merge at random with the average frequency ω. After mixing, two new particles are instantaneously produced by breaking up of the coalesced complex. The characteristic time constant is $t_{CD} = 1/\omega$. This is mainly a phenomenological parameter. In turbulent media, it is generally accepted that t_{CD} is proportional to Corrsin's time constant τ_s (See above).

Mixing by molecular diffusion. This is the ultimate and finally the only process really able to mix the components of a fluid to the molecular scale. The time constant for this process is the diffusion time $t_D = \mu L^2/\mathcal{D}$. μ is a shape factor and L is the ratio of the volume to the external surface area of the particle. For instance let us consider various shapes : slabs (thickness 2R, case of lamellar structure with striation thickness δ = 2R), long cylinders (diameter 2R, case of filamentous structure), and spheres (diameter 2R, case of spherical aggregates).

It can be shown (110) that

$$\mu = \frac{p+1}{p+3}, \quad L = \frac{R}{p+1} \text{ and hence } t_D = \frac{R^2}{(p+1)(p+3)\mathcal{D}} \qquad (7\text{-}9)$$

where p = 0 for slabs, p = 1 for cylinders and p = 2 for spheres. The particle then approximately behaves as a first order dynamic system of time constant t_D with respect to mass transfer. The problem, when molecular diffusion is retained as the mixing mechanism is to choose the adequate length scale ℓ (or 2R). Should ℓ (or 2R) be identified with the striation thickness δ or with one of the viscous dissipation microscales of Table I (Kolmogorov $\lambda_K = (\nu^3/\varepsilon)^{1/4}$, Batchelor $\lambda_B = (\nu\mathcal{D}^2/\varepsilon)^{1/4}$, Corrsin $\lambda_C = (\mathcal{D}^3/\varepsilon)^{1/4}$) or with Corrsin's microscale for segregation dissipation λ_s ($\lambda_s^2 = 24\mathcal{D}\,(L_s^2/\varepsilon)^{1/3}$? This may depend on the mechanism leading to the ultimate stage of molecular mixing. An experimental clue for the choice of the relevant microscale is the influence of viscosity which intervenes in a different manner in their expression. This point is still open to discussion. Some authors find a marked influence of viscosity and incline towards the choice of $\ell = \lambda_K$ (111-116). Others find almost no influence of viscosity and conclude in favour of λ_s (2, 30, 117, 118, 39, 109, 71, 108, 32). However, it is not quite sure that all the experiments reported in the literature are free from spurious effects (macromixing, early stages of convective or erosive mixing, etc...). Rigorous experiments, in perfectly well defined conditions should be devised to settle this problem.

As the rate of energy dissipation per unit mass ε figures in all the expressions for the microscales, the problem of its spatial repartition and of the conversion efficiency of the macroscopic dissipation is also posed here. In order to give a few orders of magnitude, the approximate values of some microscales and time constants in typical conditions are reported in Table III.

Table III

Experimental parameters : $\nu = 10^{-6}\ m^2 s^{-1}$, $\mathcal{D} = 10^{-9}\ m^2 s^{-1}$, $Sc = 10^3$		
	Case 1 : ε = 1 Watt kg^{-1} L_s = 0.3 m (large reactor)	Case 2 : ε = 10 Watts kg^{-1} L_s = 0.1 m (small reactor)
Kolmogorov λ_K	32 μm	18 μm
Corrsin λ_s	70 μm	16 μm
$\lambda_K^2/\mathcal{D} = \nu^{3/2}/(\mathcal{D}\varepsilon^{1/2})$	1 s	0.3 s
$\tau_s = 2\ (L_s^2/\varepsilon)^{1/3}$	0.9 s	0.2 s
$t_\delta = \sqrt{2}\ \tau_K = (2\nu/\varepsilon)^{1/2}$	1.4×10^{-3} s	0.45×10^{-3} s
$\lambda_C^2/\mathcal{D} = (\mathcal{D}/\varepsilon)^{1/2}$	3.5×10^{-5} s	

It is concluded from this Table that λ_K and λ_s lie between 10 and 100 μm, and that $\lambda_K^2/\mathcal{D}$ and τ_s have the same order of magnitude between 0.1 and 1 second. Other processes are much faster.

Mixing and chemical reaction. We must now discuss the coupling between the mixing processes and chemical reactions. Simulations via the IEM model or the C-D model have already been presented in the preceding section. The key parameter is the ratio of some characteristic reaction time (e.g. $t_R = 1/k\ C_o^{n-1}$ for an nth order reaction) to the micromixing time t_m (or t_{C-D}). When t_R/t_m is small, the chemical reaction is faster than mixing and microscopic concentration gradients appear: the fluid becomes segregated. Conversely, when t_R/t_m is large, the fluid may be considered as well micromixed from the viewpoint of the chemical reaction (98).

Several authors have studied the competition between chemical reactions and molecular diffusion in fluid particles. In stirred reactors, Nauman (81) introduced a segregation number $N_{Seg} = \phi R^2/(\pi^2 \mathcal{D}\ \tau)$, which compares the diffusion time with the space time. A major contribution to this problem was brought by Rys, Bourne and their group at the E.T.H. of Zurich. They developed a model (the MIRE model) essentially adapted to predict the yield of consecutive competing reactions $A + B \xrightarrow{k_1} R,\ R + B \xrightarrow{k_2} S$ in stirred reactors (119, 120). The particles are assumed to be spherical (radius R) and the equations for reaction/diffusion are solved in various cases: Version I: Particles initially containing A immersed in pure B (A is not allowed to diffuse within the particle, other species are). Version II : symmetrical case (particle of non-diffusing B immersed in A). The parameters are two Thiele criteria $\phi^2_{Bi} = k_i B_o R^2/D \sim t_D/t_{Ri}$, $E = A_o/B_o$ and $\alpha = V_A/V_B$. The model predicts the yield of S : $X_S = 2S/(2\ S + R)$ at the end of the reaction (when all B is consumed). It was developed for batch and semi-batch reactors (119, 120), and later extended to continuous stirred reactors via a somewhat complicated procedure (121-112). Some criticism may be adressed, to the MIRE-model, in spite of its great interest : arbitrary choice of spherical shape, assimilation of R to half the Kolmogorov microscale (which is not obvious as we have seen above) and above all, assumption that the initial reactant in the particle cannot diffuse outside, which creates an unwanted dissymmetry between A and B when $V_A = V_B$.

The reaction/diffusion competition can also be simulated by the IEM-Model. It suffices to set $t_m = t_D = \mu L^2/\mathcal{D}$. The equivalence with the MIRE-model will be discussed below.

More sophisticated mechanisms may be considered, where chemical reaction takes place during various stages of Beek and Miller, or a combination of these :

- Erosive mixing followed by reaction : Two unmixed reactants come into contact in a CSTR by erosion of fresh aggregates. The erosion product is either a microfluid (71) or small segregated particles of mixed reactants undergoing further interaction by molecular diffusion (108).
- Stretching of particles and simultaneous reaction.

For representing reaction and diffusion in stretching lamellar structures, Ottino and Ranz (101) introduced a "warped time" t_w defined by :

$$t_w = \int_0^t \frac{\mathcal{D}}{\delta(t')^2} dt' \text{ or } \frac{dt_w}{dt} = \mathcal{D} a_v^2 \qquad (7\text{-}10)$$

where the rate of change of the striation thickness is given by (7-2). The equations for reaction/diffusion are then easier to solve in Lagrangian coordinates. In the same way Bourne, Angst et al. (122, 114) extended their MIRE-Model by assuming that the size of the particles decreases according to (7-5). This would mean that a reduction of the striation thickness below the Kolmogorov microscale is conceivable. Reaction and diffusion in distorting structures were also studied by Palepu et al. (31) and Spalding (123). The I.E.M. Model accounts for such deformations in assuming that the micromixing time is a specified function of the particle age (32).

We may now complete the list of equivalences cited in (6-9) by a new one between the droplet-diffusion model of Nauman (81) or the MIRE-model of Rys, Bourne et al. (119-120), and the IEM Model. The condition for equivalence is $t_m = t_D$, where t_D is given by (7-9). The IEM-Model is a lumped version of the distributed parameter reaction/diffusion model. Numerical simulations prove that replacing concentration profiles by average values does not change the overall conversion and yield in the particle very much, even for "stiff systems" of fast reactions. Figures 11 and 12 are examples of such equivalences. Therefore, my opinion is that in most applications, a simple lumped parameter model may be used in place of sophisticated distributed models. This saves much computer time and the difference in the results is not greater than for instance that induced by a change in the arbitrary assumptions concerning the particle shape.

The simulations via the IEM model revealed an interesting property. A partially segregated fluid may be considered as a mixture of macrofluid (fraction β) and microfluid (fraction $1 - \beta$). It comes out that the ratio $(1-\beta)/\beta$ is always close to that of two characteristic times (32). In the case of erosive mixing of two reactants in a CSTR $(1 - \beta_e)/\beta_e \approx \tau/t_m \approx 4\ \tau/t_e$.

In the case of reaction and diffusion of premixed reactants in a CSTR (figure 13) : $(1-\beta_D)/\beta_D \approx t_R/t_D$. This is an interesting rule of thumb for rapid estimation of the extent of segregation.

Identification of segregation by chemical methods. Partial segregation can be studied through its influence on the conversion and yield of chemical reactions. For instance, let us denote by X_{macro} and X_{micro} the limiting extents of reaction one would observe in a well macromixed reactor. If the reactor is partially segregated : $X = \beta X_{macro} + (1 - \beta) X_{micro}$.

This also holds for the yield of an intermediate product and is the basis for the determination of β or conversely, for the prediction of X (32). Fast consecutive-competing reactions $A + B \xrightarrow{k_1} R$, $R + B \xrightarrow{k_2} S$ ($k_1 >> k_2$) are especially interesting : if a small amount of B is mixed into an excess of A, R is formed and immedi-

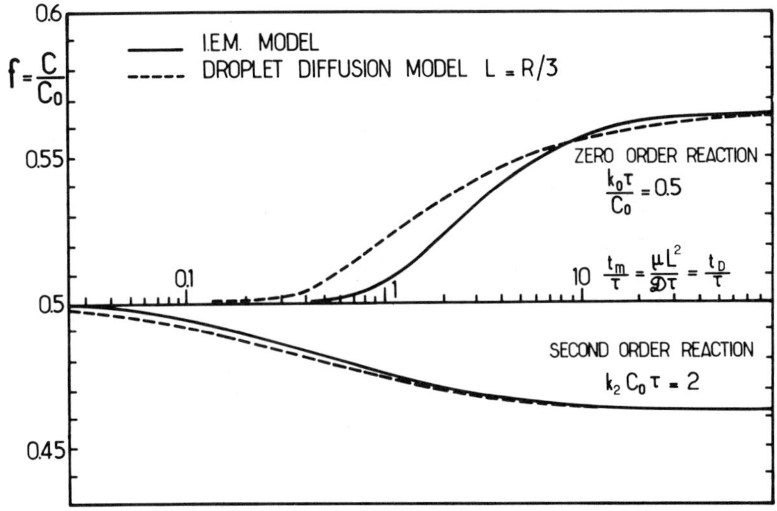

Figure 11. Equivalence between the droplet diffusion model (81) and the IEM model for a zero-order reaction and a second-order reaction in a CSTR. The Damköhler numbers are such that f = 0.5 for perfect micromixing. The agreement is excellent for the second-order reaction, more approximate for the zero-order one.

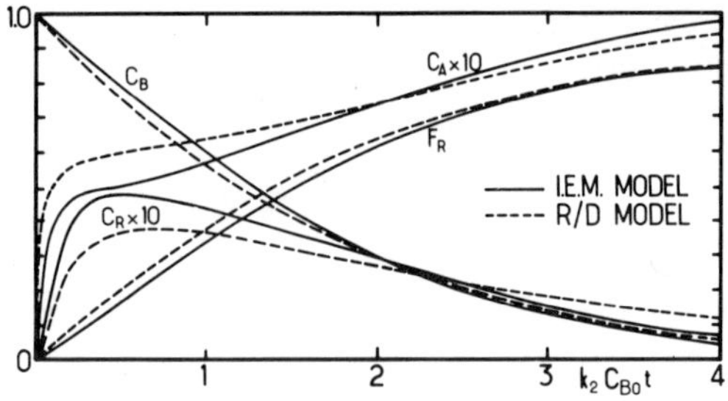

Figure 12. Equivalence between the reaction/diffusion model and the IEM model for second-order consecutive competing reactions $A + B \xrightarrow{k_1} R$, $R + B \xrightarrow{k_2} S$. The curves represent the average concentrations vs. (reduced) time in a spherical particle immersed in a bath of constant composition ($C_A = 0.105$, $C_R = 0$). Initial concentrations in the particle: $C_{Ao} = 0$, $C_{Bo} = 1$, $C_{Ro} = 0$. $k_2 C_{Bo} R^2/\mathcal{D} = 2$ for A and R. B cannot diffuse within the particle ($\mathcal{D} = 0$). $k_1/k_2 = 10$. F_R represents the total production of R (equivalent concentration in the particle). Basis for the equivalence $t_{mj} - t_{Dj} = \mathcal{D}L^2/ j$; $\mu = 3/5$, $L = R/3$ (sphere). The agreement is satisfactory. In much "stiffer" conditions ($k_1 >> k_2$), the agreement is always good for the overall production of R, even if the individual concentration profiles become different.

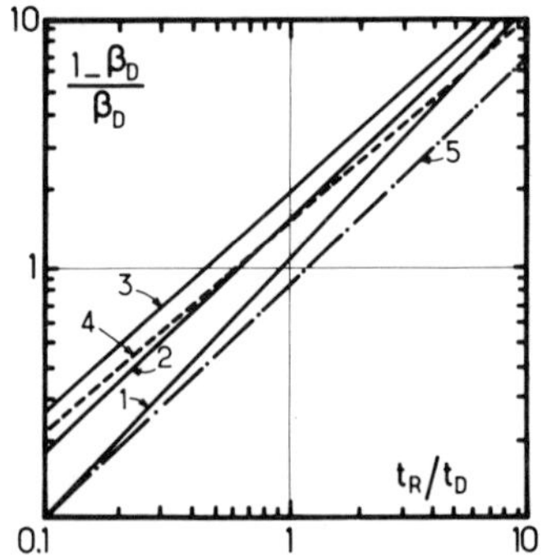

Figure 13. Microfluid/macrofluid volume ratio vs. reaction/diffusion time ratio. Key to curves: 1 to 4, simulation with the IEM model, $t_m = t_D$; 1 to 3, second-order reaction $k_2 C_{Ao} = 2(1), 5(2), 10(3)$; 4, second-order consecutive competing reactions $C_{Ao} = C_{Bo}$, $k_1/k_2 = 2$, $t_R = 1/k_1 C_{Ao}$; 5, reaction and diffusion in a slab (See Ref. 32.).

ately converted to S at the contact of B if the fluid is segregated. Therefore, X_s is some kind of segregation index : $X_{s,micro} < X_s < 1$, and $\beta = (X_s - X_{s,micro})/(1 - X_{s,micro})$.

The method was extensively exploited by Bourne and coworkers, who enumerated the qualities of a good reaction for industrial tests (124)($t_R \approx t_D$, distribution of products depending on segregation, irreversible reaction, known mechanism, easy analysis and regeneration, safe and inexpensive chemicals). Unfortunately, the reactions proposed to-date do not fulfill all these requirements : Azo-coupling (119, 137), nitration of aromatic hydrocarbons (125), bromation of resorcin (124), bromation of 1-3-5 trimethoxybenzene (126). Some of them exhibit complex pH effects (124, 127, 111), but also a good sensitivity to segregation. For instance, the amount of 2-4 dibromoresorcin in the di-isomer may vary from 30% to 60 % when the agitation speed passes from 0 to 360 r.p.m. (124). A new reaction is proposed in this Symposium by Barthole et al. (128) : the precipitation of baryum sulfate complexed by EDTA in alkaline medium under the influence of an acid. It has many of the advantages cited above, but the reaction is not strictly instantaneous and thus dependent on macromixing. Fluorescence methods may also be employed (20). In a recent series of papers (129, 112, 113, 114), Bourne and coworkers presented a thorough investigation of segregation in stirred reactors (2.5 and 63 dm^3) with various inlet positions. As a test reaction they used the coupling of 1-naphtol (A) with diazotised sulphanilic acid (B). They succeeded in applying the MIRE-Model provided that the particle size 2R be 2 to 9 times smaller than the Kolmogorov microscale. The agreement was improved by assuming stretching of particles : $\delta = \delta_o (1 + t^2/t_\delta^2)^{-1/2}$ where $t_\delta = 2 (\nu/\varepsilon)^{1/2}$ (122, 114). They showed that the best mixing conditions were achieved with inlet tubes placed just beneath the turbine. They concluded from their interpretation, relying on the Kolmogorov microscale, that the scale up of the state of segregation requires keeping ε and thus N^3d^2 constant. The interpretation of these beautiful experiments should be discussed with much care, taking into account possible macromixing effects (comparison of reaction time and circulation time t_c), intervention of different mixing processes (of stage 2) preceding the diffusional one, and the spatial distribution of ε in the tank.

Beside consecutive-competing reactions, instantaneous (generally acid-base) reactions are also used as an indicator of segregation, especially in multijet tubular reactors. Ottino (102) deduced the relationship $t_w(t)$ between "warped" and real time from the comparison between experimental conversion X(t) along the axis of a tube and the theoretical expression $X(t_w)$. a_v was then calculated by (7-10) and the efficiency eff(t) by : $(1/a_v)da_v/dt = \mathrm{eff}(t)(\underline{D} : \underline{D})^{1/2}$. eff(t) was found to decrease as a function of t with $\delta = 1/a_v$ of the order of 10 μm. In the same paper, the average efficiency "eff" was estimated in a stirred reactor. "eff" was found to decrease from 20-30 % for t < 0.25 s to 1-10 % later, but here also, macromixing effects (circulation time ?) cannot be excluded. Bourne et al. (116) also performed experiments with NaOH/HCl mixing behind a Sulzer static mixer in a tube. ε was estimated by $U(\Delta p/\ell)/\rho$ and the influence of the viscosity was especially studied, and found to be noticeable. Assuming a diffusional mechanism for mixing, these authors conclude that Corrsin's micromixing time τ_s cannot explain their observations and that the influence of viscosity suggests a diffusion time based on Kolmogorov microscale. The same conclusion was drawn from experiments in stirred tanks, where the IEM-model seemed unable to account for all experimental data. This is in contradiction with the results of Pohorecki and Baldyga (117) who also studied the reaction of NaOH and HCl controlled by mixing in a tube. They found that the IEM-model was applicable with $t_m = 2\ \tau_s \sim (L_s^2/\varepsilon)^{1/3}$ where L_s was equal to the diameter of the injection tube. ε was deduced from the assumption of isotropic homogeneous turbulence.

This problem of mixing with chemical reaction has drawn the attention of many authors in the last few years. Takao et al. (130) studied the alkaline hydrolysis of chloroacetate in a batch stirred reactor ($t_R \not\approx t_m$) ; their results, obtained on the basis of the IEM model, can probably be explained by macromixing ($t_m \sim t_c \sim 1/N$). Miyawaki et al. (118) studied the reaction of $CO_2 + NH_3$ and $CO_2 + OH^-$ in multijet tubes and in stirred reactors. In tubes, the data are compatible with the IEM model ($X = 1 - \exp(-t/t_m)$) with $t_m \sim$ Corrsin's τ_s, whereas in stirred tanks conversion is probably controlled by recirculation ($t_m \sim 1/N$) as in reference (130). Murakami et al. (131) developed a model equivalent to the IEM model for interpreting mixing in batch stirred reactors (1 and 50 ℓ) both with a non reactive tracer and in the presence of three reactions of different rapidity. They found that Nt_m could be correlated as a function of the agitation Reynolds number Nd^2/ν and a Damköhler number. Costa and Lodi (136) assimilated the IEM mixing time t_m to $(\nu/\varepsilon)^{1/2}$ with a correction depending on the Schmidt number, but without any experimental support. Hanley and Call (132) suggested exploiting concentration fluctuations at the outlet of a CSTR to calculate micromixing parameters. Ghodsizadeh and Adler (133) proposed an interesting method based on dilatometry to follow the course of an acid-base reaction in a batch reactor. Bhatt and Ziegler (134) determined the macrofluid fraction in a CSTR by taking into account the non-ideality of the segregated phase RTD and by assuming interaction by reaction and diffusion. Bryant (135) considered the case of zero order reaction in fermenters by assuming diffusional limitations in particles having the Kolmogorov size λ_K. Palepu et al. (31) used the "warped"-time method described in (102) to estimate the striation thickness δ as a function of time in a multijet tube and in a stirred reactor. After an initial decrease, it seems that δ tends to λ_K but there is still some strange behaviour (climb of δ before stabilization) and the problem of the efficiency in the calculation of ε is also posed. From their own experiments (109, 71, 108), Villermaux and coworkers suggest (32) that depending on experimental conditions, and chemical reactions, there are several mixing mechanisms for stages 2 and 3 of Beek and Miller occuring simultaneously and interacting with chemical processes. This might explain the discrepancies observed between authors willing to interpret their experiments by one single mechanism.

Conclusion. There are still uncertainties in the final interpretation of mixing and chemical reaction at the molecular level. The IEM model seems to provide a simple basis for representing interaction between particles, even by molecular diffusion. The problem is to decide what is hidden behind the micromixing time t_m ? Corrsin's time constant τ_s (32) ? A diffusion constant based on Kolmogorov microscale (113, 114) ? Further research should be developed in the following directions :

- Search for reliable test reactions obeying the criteria stated in reference (124). These reactions should be usable in industrial reactors and perhaps be less "stiff" than those proposed by Bourne and coworkers.
- Design of experiments where hydrodynamic conditions are perfectly controlled : small stirred reactors with high power input (no macromixing effects, circulation time $t_c \ll t_R$, no spatial variation of ε), multijet tubular reactors with a diameter large enough for pure homogeneous isotropic turbulence to be achieved.
- Clear distinction between the mechanisms pertaining to successive or simultaneous stages of mixing.
- Once the above requirements are fulfilled, a systematic study of the influence of viscosity ν and power dissipation ε should be made.

The Stirred Tank

A review on mixing in chemical reactors would be incomplete without an account of recent progress in the understanding of the working of stirred tanks, which remain the basic devices for industrial mixing.

Internal circulation patterns and turbulence. Sasakura et al. (138) and Rachez et al. (139) investigated internal age distributions by means of tracers and proposed representative models of the circulation pattern. Bryant and coworkers (135, 140, 141) and Reuss studied more specifically circulation times in large stirred tanks by using radio flow-followers (aiming at applications for fermenters > 0.5 m^3). They showed that circulation times t_c were log-normally distributed :

$$f(t_c) = \frac{1}{\sqrt{2\pi}\sigma t_c} \exp\left[-\left(\frac{\ln t_c - \mu}{\sigma\sqrt{2}}\right)^2\right] \qquad (8\text{-}1)$$

the mean value $\overline{t_c}$ and natural variance s^2 being given by

$\overline{t_c} = \exp(\mu + \sigma^2/2)$ and $s^2 = (\overline{t_c})^2(\exp \sigma^2 - 1)$. If H is the height of liquid in the tank, they found $\overline{t_c} \sim H/(Nd^3)$ and $s \sim H^{7/3}/(Nd^3)$. They also studied the terminal mixing time θ_m, required for reducing concentration gradients down to a specified level by multiple recirculations. They found $\theta_m/\overline{t_c} = A + B(s/\overline{t_c})^2$, $s/\overline{t_c} > 0.8$, so that $\theta_m \sim H^{11/3}/(Nd^3)$ when A is small. The power input is thus $P \approx (Nd)^2(V/\overline{t_c})$.

Many correlations for mixing time (see above) have been proposed in the literature (142). One of the most comprehensive treatments of this problem was published by Khang and Levenspiel (143), on the basis of a recycle model. θ_m is defined as the time constant for the exponential decrease of pseudo-periodic oscillations after a pulse injection of tracer in a batch stirred reactor. When Re > 10^4, they obtain for turbines :

$$N\,\theta_m(d/d_T)^{2.3} = 0.5 \approx 0.1\;P/(\rho N^3 d^5) \quad (8\text{-}2)$$

and for propellers

$$N\theta_m\,(d/d_T)^2 = 0.9 \approx 1.5\;P/(\rho N^3 d^5) \quad (8\text{-}3)$$

Correlations for jet-mixing can be found in (157).

Experimental data on mixing times may be used to estimate the overall efficiency for batch mixing of viscous fluids, according to the method proposed by Ottino et al. (107).

In addition to these macromixing characteristics, many authors have determined turbulence parameters and their spatial distribution within the tank volume by measuring velocity and concentration fluctuations (144-147, 19, 158). In a typical investigation (19) concerning a semi-industrial tank (0.15 m^3) and aqueous medium, the following spatial variations were found : $u' = 5$ to 30 % of πNd, $L_f = 4$ to 150 mm, $\lambda_f = 1$ to 5 mm, $\varepsilon/\overline{\varepsilon} = 0.2$ to 2.5, $c'/\overline{C} = 2$ to 10×10^{-4} (for eddies > 100 μm). This shows that a stirred tank is far from being the homogeneous and uniform system assumed in many academic papers.

Cell models. In order to predict chemical conversion in stirred tanks, Patterson and coworkers (3, 39, 40) divided the tank volume into 30 mixing segments connected by specified flowrates $Q_{ij} \sim Nd^3$. The turbulence level in each segment is characterized by L_s ($\sim d_T$) and $\varepsilon(\sim d_T^2)$ (HDM model). Mann and coworkers (148,149) also studied a model where cells (or segments) are connected according to the average flow pattern. Commutation according to a specified probability at each cell's outlet allows a stochastic path to be simulated, for instance for a flow follower. They thus obtained circulation time distributions very similar to experimental ones (135, 140, 141).

Multiphase stirred tanks. This item will be reviewed only very briefly as the subject was recently covered in excellent and quasi-exhaustive surveys by Joshi et al. for gas-liquid reactors (150) and Tavlarides and Stamatoudis for liquid-liquid reactors (5, 4). Joshi et al. gave a thorough comparison of correlations for $N\theta_m$. This product seems to increase with the gas-flowrate. Little is yet known about the state of mixing of the dispersed gas and on the influence of solid in suspension.

Tavlarides presents a sophisticated model for representing coalescence and breakage of droplets in liquid-liquid dispersions. The model relies on the population balance equation and still requires the adjustment of 6 parameters. The solution of such equations is difficult and requires the use of Monte-Carlo methods (151). The effect of coalescence and break-up of droplets on the yield of chemical reactions was studied by Villermaux (33). Micromixing effects may occur even in batch reactors if there is a drop size distribution and mass-transfer control. Although practical rules for the design and scale-up of liquid-liquid reactors are available as Oldshue showed in the case of alkylation (152), many problems remain unsolved (5) : mass transfer effects, high hold-up fractions (> 20 %), large density differences, high viscosities, influence of surfactants.

Conclusion : areas for future research. Mixing in stirred reactors is no longer the empirical operation it used to be, ("mostly art and very little science" (153). For instance, Oldshue summarized useful rules for the scale-up of fermenters (153). However, several current problems are still waiting solution. These were reviewed in an excellent paper by Kipke (154). Future research should be directed towards

- Turbulent phenomena
- Large volumes (spatial unhomogeneities)
- Multiphase systems (gas-liquid, liquid-solid, liquid-liquid, gas-liquid-solid)
- Non newtonien media, rheology problems (155, 156)
- Search for simplified models and new concepts
- Less dimensional analysis

Less classical devices : static mixers

Static mixers have been essentially developed since 1970. About 30 types of these devices are known (159). Their effectiveness can be characterized in two ways : by the reduction of $\sigma/\overline{C}$ (See above) along the mixer axis (159), $\sigma/\overline{C} = a \exp(-mz/d_t)$ or by the increase of striation number produced by passing through n mixing elements (160) : $\delta_o/\delta = b^n$.

For instance, with the Hi-Mixer (161), $\delta_o/\delta = 4^n$. A factor of 5 is easy to obtain for m with respect to an empty tube, at the expense of a corresponding increase in pressure-drop. A comparison between the existing types of static mixers (162) shows that most of them have an equivalent effectiveness. The case of Sulzer-Mixers has been especially studied (163), including use with gases (164).

Mixing in static mixers considered as chemical reactors was essentially studied by Nauman (165, 166). This author proposed a model which consists of a tubular reactor comprising N zones in laminar flow (parabolic velocity profile). Mixing between each zone is achieved accross a plane by a permutation of the radial position of fluid particles ($r_1 \rightleftarrows r_2$), in this way the flowrate is kept unchanged . Several cases are considered : complete mixing (permutation at random), complete flow inversion ($r_2 = 1 - r_1$, $W_2 = 1 - W_1$), partial inversion. In the first case, N = 0 corresponds to a CSTR and $N \to \infty$ to a plug-flow reactor. It is shown that the best chemical conversion is obtained with complete flow inversion. The RTD in a Kenics mixer comprising 8 elements could be represented by this model with N = 3 and complete mixing. Static mixers could be used as chemical reactors for specific applications (reactants having large viscosity differences, polymerizations) but the published data are still very scarce and additional information is required for assessing these possibilities.

Beside static mixers, there are practically no alternatives to the "ubiquitous" stirred tank, if one excepts loop reactors (167) and the somewhat special back-flow mixer (168). Imagining entirely new principles for mixing reactants is a challenge for future researchers. First estimations show that an "informed" mixing system, working as a Maxwell demon would be much more effective than our present devices (169).

Mixing of Solids. This point is actually very important but deserves a special review, and will not be treated in this paper. The reader will find the state of the art and a literature survey in three excellent papers (170, 171, 172).

An example of the importance of mixing effects in chemical reactors : continuous free radical polymerization. One might now ask the question : are segregation effects really important in practice or is micromixing "a solution in search of a problem" (173), a mere intellectual exercise for academics who are short of original PhD subjects ?

It is true that micromixing effects can generally be neglected in the design of reactors for simple and slow reactions. However, as has been pointed out in the preceding Sections in the case of fast reactions with unmixed reactants, chemical conversion could be entirely controlled by mixing, and induce dramatic variations in the distribution of products. The practical examples of combustion and reactions in liquid suspension are especially illuminating in this respect.

Another area where micromixing plays a cardinal role is continuous polymerization.

The subject is treated elsewhere in this Symposium, and was reviewed by Nauman (173) and Gerrens (174) a few years ago. Therefore a thorough discussion of mixing effects in polymer reactors would go beyond the scope of this paper. It is likely that significant progress has been made since these reviews, but they are difficult to assess because of a lack of published data. Two examples will show the importance of micromixing effects on the structural characteristics of polymers (175). The first example is a simulation of free radical polymerization in a CSTR. The retained kinetic scheme is

$$A \xrightarrow{k_d, f} 2R \qquad \text{Initiation}$$

$$R + M \xrightarrow{k_p} R \qquad \text{Propagation}$$

$$R + P \xrightarrow{k_{tP}} P + R \qquad \text{Transfer to Polymer}$$

$$R + R \xrightarrow{k_t} P + P \qquad \text{Termination (Disporportionation)}$$

Figure 14 shows a plot of the Dispersion Index DI = $\overline{M}_w/\overline{M}_n$ versus the conversion X of the monomer for segregated flow (S) and well micromixed flow (M). The dramatic influence of segregation can be noticed at high conversion, especially with transfer to polymer. Moreover, an interesting effect is observed with diluted and slow initiators, namely an inversion of the relative position of S and M curves when the transfer constant k_{tP} is increased. This doesn't happen with concentrated and fast initiator. The second example is an experimental one (176). Continuous polymerization of styrene was carried out in a CSTR and in cyclohexane solution in order to keep the viscosity low and constant. The Dispersion Index was measured as a function of space time and agitation speed. Limiting curves for segregated flow (S) and well micromixed flow (M) were calculated from batch experiments. Clear evidence for segregation effects can be seen on figure 15 which shows that perfect micromixing may be very difficult to achieve, even with strong agitation and low viscosity.

Besides these laboratory experiments, the analysis of industrial reactors may also reveal segregation effects, as for instance in reactors for free radical polymerization of ethylene where the initiator feedstream is likely to be mixed by an erosive process (175). Polymerization and polycondensation reactors offers an especially interesting field for future applications of micromixing.

General conclusion

The end of this survey leaves us with the feeling that research on mixing in chemical reactors is a very lively area, where problems have been attacked from several directions (turbulence theory, RTD and mixing earliness, segregation and micromixing ...). If the major concepts have been identified, there is still a need for a unified theory allowing a-priori predictions from the sole knowledge of physicochemical properties and operating parameters, even if encouraging progress has been made in this direction.

Without repeating the conclusions drawn at the end of each

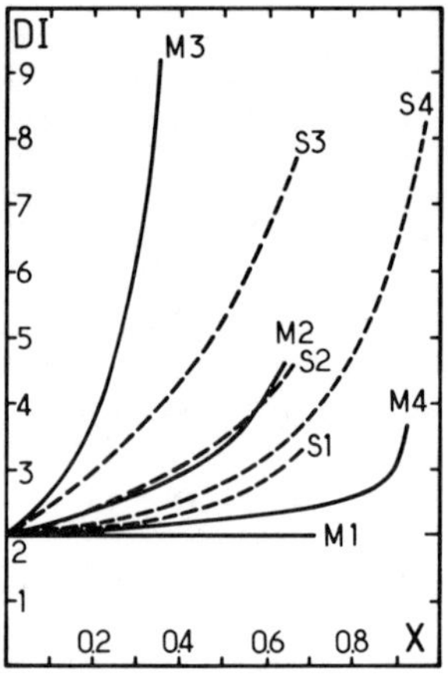

Figure 14. Free radical polymerization in a CSTR. Dispersion Index DI vs. conversion X. S - segregated flow, M = well-micromixed flow, I = linear polymerization $k_p = 5 \times 10^7$ L mol^{-1} h^{-1}, $k_t = 1.5 \times 10^{12}$ L mol^{-1} h^{-1}, $k_d = 0.033$ h^{-1}, f = 0.5, $A_o = 3 \times 10^{-3}$ mol L^{-1}, $S_o = 7.12$ mol L^{-1} (solvent), $M_o = 3.56$ mol L^{-1}. Curves 2 to 4: Transfer to polymer. 2 : $k_{tP} = 3.5 \times 10^3$ mol L^{-1} h^{-1}, 3 : $k_{tP} = 1.05 \times 10^4$ L^{-1} mol^{-1} h^{-1}, 4 : $k_{tP} = 1.05 \times 10^4$ L^{-1} mol^{-1} h^{-1}, $A_o = 3 \times 10^{-2}$ mol L^{-1}, $k_d = 0.33$ h^{-1} (other parameters unchanged).

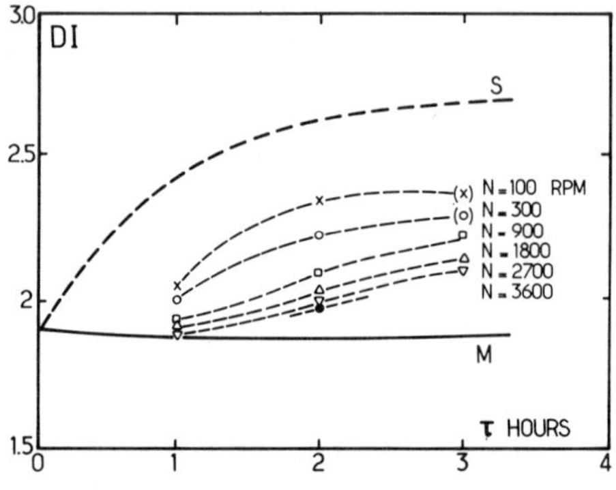

Figure 15. Effect of segregation on polymerization of styrene in cyclohexane solution. Standard CSTR with 4 baffles and a 6-blade turbine, V = 670 cm^3, T = 75 $^{\circ}$C. Dispersion Index DI vs. space time. Influence of agitation speed. Curves S (segregated flow) and M (well-micromixed flow) calculated from batch experiments. Initiator: PERKADOX 16, $A_o = 0.033$ mol L^{-1}, $k_d = 5 \times 10^{-5}$ s^{-1}, f = 0.85; $M_o = 6.65$ mol L^{-1}, $S_o = 2.22$ mol L^{-1}.

Section, we may try to summarize the situation as follows : Residence time distributions are now a well established tool but progress is still desirable for multiple inlet/outlet systems, unsteady state operations, fluids with time-variable properties and inhomogeneous reacting media. More attention should be paid to internal age distributions and related quantities such as circulation time distributions in stirred tanks. In continuous reactors, models for mixing earliness describing the transfer between Entering and Leaving Environments are superabundant. Further reduction of segregation by interaction between fluid particles can be conveniently represented by simple models (exchange with the mean, coalescence-dispersion), but several stages for mixing, each with their own time constants should be considered, possibly in series or in parallel. There is still a problem as to the ultimate stage of mixing by molecular diffusion where it is not clear whether the mixing time is $t_m = \tau_s \sim (L_s^2/\varepsilon)^{1/3}$ or $t_m = t_D \sim \lambda_K^2/\mathcal{D}$. Carefully designed experiments (no macromixing effects, perfectly defined hydrodynamic patterns) and new chemical test reactions would be welcome in this respect. The method of "characteristic times" is especially helpful for determining which processes are controlling. These are for instance the space time τ for a continuous reactor ; a characteristic time for internal macromixing pattern, e.g. the circulation time t_c in a stirred tank ; one or several reaction times, e.g. $t_R = 1/kC_o^{n-1}$; and one or several micromixing times t_m, e.g. t_e (erosion) or $t_D \sim \ell^2/\mathcal{D}$, or $t_{C-D} = 1/\omega$ or $t_\delta \sim (\nu/\varepsilon)^{1/2}$, or $\tau_s \sim (L_s^2/\varepsilon)^{1/3}$. Comparison between all these times allows the determination of the mixing regime, sometimes quantitatively. It was thus established that the microfluid/macrofluid volume ratio was nearly equal to t_R/t_D. Another characteristic of current research is a gradual and fortunate merging between the Eulerian approach of Fluid Mechanics and the Lagrangian approach of Chemical Engineering. Measurement of concentration fluctuations should be developed, both in presence and in absence of chemical reactions in order to obtain reliable spectral data. However, the final solution to micromixing problems should not be sought in turbulence theory alone, but rather in phenomenological interaction models, whose parameters could have a fundamental interpretation by this theory.

This is the wish of most industrials : "In general, industry would plead for less sophisticated mathematical models and more phenomenological models giving us more understanding of what's going on in the tank" (10). Among other recommendations (9, 10, 27, 154), there is a general agreement for encouraging research on large volume reactors, gas-liquid-solid systems and mixing of non newtonian fluids. New ideas on entirely novel mixing principles and equipments would also be welcome. But above all, theory will progress in a direction useful to practitioners if more experimental data on realistic industrial situations are available to researchers.

Literature Cited

1. Nauman, E.B. Chem. Eng. Commun. 1981, 8, 53-131
2. Brodkey, R.S. Chem. Eng. Commun. 1981, 8, 1-23
3. Patterson, G.K. Chem. Eng. Commun. 1981, 8, 25-52
4. Tavlarides, L.L. Chem. Eng. Commun. 1981, 8, 133-164
5. Tavlarides, L.L.; Stamatoudis, M. Adv. Chem. Eng. 1981, 11, 199-273
6. Ritchie, B.W.; Tobgy; A.H. Chem. Eng. Comm. 1978, 2, 249-264
7. Cheng, D.C.H.; Tookey, D.J. Proceed. Second Europ. Conf. Mixing, Cambridge 1977, BHRA Fluid Engineering
8. Paul, E.L.; Diena, J.R.; Nusim, S.H.; Sklarz, W.A. AIChE Meeting 9-12, 1981, New Orleans, Louisiana, Paper 18f.
9. Adler, R.J.; National Science Foundation Workshop, August 21-26, 1977, Rindge, New Hampshire
10. Short, D.G.R.; Etchells, A.W. Proceedings of the Fourth Europ. Conference on Mixing, April 27-29 1982, (BHRA Fluid Engineering), Paper A1, 1-10
11. Danckwerts, P.V. "Insights into Chemical Engineering. Selected papers of P.V. Danckwerts"; Pergamon Press, Oxford 1981
12. Danckwerts, P.V. Chem. Eng. Sci. 1958, 8, 93-99
13. Zwietering, T.N. Chem. Eng. Sci. 1959, 11, 1-15
14. Danckwerts, P.V. Appl. Sci. Research 1952, 3, 279-298
15. Nagata, S. "Mixing. Principles and applications"; Kodansha Ltd. Tokyo, 1975
16. Brodkey, R.S. "Application of Turbulence Theory to Mixing Operations"; Acad. Press, New-York, 1975
17. Hiby, J.W. Internat. Chem. Eng. 1981, 21, 197-204
18. Brodberger, J.F. These, Inst. Nat. Polytech. Lorraine, 1981
19 Barthole, J.P.; Maisonneuve, J.; Gence, J.N.; David, R.; Mathieu, J.; Villermaux, J. Chem. Eng. Fund., 1982, 1, 17-26
20. Patterson, G.K.; Bockelman, W.; Quigley, J. Proceedings of the Fourth Europ. Conf. on Mixing, April 27-29, 1982, (BHRA Fluid Engineering), Paper J1, 303-312
21. Käppel, M. VDI Forschungsheft 578, 1976
22. Käppel, M. Internat. Chemical Eng. 1979, 19, 571-590
23. Lehtola, S.; Kuoppamäki, R. Chem. Eng. Sci. 1982, 37, 185-191
24. Edwards, W.M.; Zuniga-Chaves, J.E.; Worley, F.L.; Luss, D. Ind. Eng. Chem. Fundam. 1976, 15, 341
25. Ogawa, K.; Ito, S. J. Chem. Eng. Japan 1975, 8, 148-151
26. Ogawa, K.; Ito, S.; Matsumura, Y. J. of Chem. Eng. Japan 1980, 13, 324-326
27. Brodkey, R.S. AIChE 1981 Annual Meeting, New Orleans Nov 8-12, Paper n° 61c
28. Bennani, A.; Alcaraz, E.; Mathieu, J. C.R. Acad. Sc. Paris 1981, 293, Ser. II; 641-644
29. Bennani, A.; Alcatraz E.; Mathieu, J. C.R. Acad. Sci. Paris 1981, 293, Ser. II, 739-741
30. Bennani, A.; Gence, J.N.; Mathieu, J. C.R. Acad. Sci. Paris 1981, 293, Ser. II, 791-794
31. Palepre, P.T.; Adler, R.J.; Edwards, R.V. 74th Annual AIChE Meeting, New Orleans, Nov. 8-12, 1981, paper 18c.
32. Villermaux, J.; David, R. Chem. Eng. Commun. in press and Proceedings 2nd World Congress Chem. Eng. Montreal 1981 Oct 4-9, 4, 397-401
33. Villermaux, J. Multiphase chemical reactors. Vol. I Fundamen-

tals, A.E. Rodrigues, J.M. Calo and N.H. Sweed Editors, Nato Adv. Study Inst. Series, E-N° 51, Sijthoff and Noordhoff, 1981, 285-362
34. Mao, K.W.; Toor, H.L. Ind. Eng. Chem. Fund. 1971, 10, 192-97
35. Vassiliatos, G.; Toor, H.L. AIChE J. 1965, 11, 666-673
36. Bennani, A.; Mathieu, J. C.R. Acad Sci. Paris 1982, 294, Ser. II, 933-936
37. McKelvey, K.N.; Yieh, H.N.; Zakanycz, S.; Brodkey, R.S. AIChE J., 1975, 21, 1165-1176
38. Berker, A.; Whitaker, S. Chem. Eng. Sci. 1978, 33, 889-895
39. Canon, R.M.; Wall, K.W.; Smith, A.W.; Patterson, G.K. Chem. Eng. Sci. 1977, 32, 1349-1352
40. Waggoner, R.C.; Patterson, G.K. ISA Transactions, 1975, 14, 331-339
41. Bourne, J.R.; Toor, H.L. AIChE J. 1977, 23, 602-604
42. Vidal,C.; Roux, J.C.; Rossi, A.; Bachelart, S. C.R. Acad. Sc. Paris 1979, t. 289, Série C, 73-76
43. Marek, M.; Havlicek, J.; Ulcek, J. Proceedings of the fourth europ. Conf. on Mixing, April 27-29 1982, (BHRA Fluid Engineering) 339-354
44. Humphrey, J.A.C. Chem. Eng. Commun. 1980, 7, 345-361
45. Tarbell, J.M.; Petty, C.A. Chem. Eng. Sci. 1977, 32, 1177-88
46. Pratt, D.T. Prog. Energy Combust. Sci. 1976, 1, 73-86
47. Gouldin, F.C. Comb. Sci. and Technol. 1973, 7, 33-45
48. Bilger, R.W. Comb. Sci. and Technol. 1979, 19, 89-93
49. Nelson, H.F. Letters Heat and Mass Transfer 1979, 6, 23-33
50. Spalding, D.B. Combust. Sci. and Technol. 1976, 13, 3-25
51. Lockwood, F.C. Combustion and Flame 1977, 29, 111-122
52. Lockwood, F.C.; Syed, S.A. Comb. Sci. and Technol. 1979, 19, 129-140
53. Pratt, D.T. J. Energy 1979, 3, 177-180
54. Hébrard, P.; Magre, P. 54th AGARD/PEP Conf. Proceedings, Oct. 1979, Cologne (Allemagne), N° 275
55. Wen, C.Y.; Fan, L.T. "Models for flow systems and chemical reactors" Dekker, New-York, 1975
56. Villermaux, J. "Génie de la Réaction Chimique. Conception et fonctionnement des réacteurs", Technique et Documentation, Paris, 1982
57. Nauman, E.B. Chem. Eng. Sci. 1981, 36, 957-966
58. Roth, D.P.; Basaran, V.; Seagrave, R.C. Ind. Eng. Chem. Fundam. 1979, 18, 376-383
59. Rautenbach, R.; Waschmann, M.; Van Gilse, J. Chem. Ing. Technik 1981, 53, 726-728
60. Treleaven, C.R.; Tobgy, A.H. Chem. Eng. Sci. 1971, 26, 1259
61. Ritchie, B.W.; Tobgy, A.H. Ind. Eng. Chem. Fund. 1978, 17, 287
62. Buffham, B.A.; Kropholler, H.W. Ind. Eng. Chem. Fundam. 1981 20, 102-104
63. Bourne, J.R.; Giger, G.K.; Richarz, W.; Riesen, W. The Chem. Eng. Journal 1976, 12, 159-163
64. Conochie, D.S.; Gray, N.B. Chem. Eng. Sci. 1978, 33, 619-621
65. Quraishi, M.S.; Fahidy, T.Z. Chem. Eng. Sci. 1982, 37, 775
66. Castellana, F.S.; Spencer, J.L.; Cartolano, A. Ind. Eng. Chem. Fundam. 1980, 19, 222-225
67. Chen, M.S.K. Chem. Eng. Sci. 1971, 26, 17-28
68. Spencer, J.L.; Lunt, R.; Leshaw, S.A. Ind. Eng. Chem. Fundam. 1980, 19, 135-141

69. Weinstein, H.; Adler, J. Chem. Eng. Sci. 1967, 22, 65-75
70. Villermaux, J.; Zoulalian, A. Chem. Eng. Sci. 1969, 24, 1513
71. Plasari, E.; David, R.; Villermaux, J. A.C.S. Symp. Series, Chem. React. Eng. Houston 1978, 65, 126-139
72. NG, D.Y.C.; Rippin, D.W.T. 3rd Europ. Symp. Chem. React. Eng. Amsterdam 1964, Pergamon Press, Oxford 1965, 161-165
73. Valderrama, J.L.; Gordon, A. Chem. Eng. Sci. 1979, 34, 1097
74. Valderrama, J.L.; Gordon, A. Chem. Eng. Sci. 1981, 36, 839-44
75. Chien-Ping Chai; Valderrama, J.L. Chem. Eng. Sci. 1982, 37, 494-496
76. Ritchie, B.W. Chem. Eng. Sci. 1982, 37, 800
77. Treleaven, C.R.; Tobgy, A.H. Chem. Eng. Sci. 1972, 27, 1497
78. Ritchie, B.W. The Canad. J. of Chem. Eng. 1980, 58, 626-633
79. Ritchie, B.W.; Tobgy, A.H. The Canad. J. Chem. Eng. 1977, 55, 480-483
80. Ritchie, B.W.; Tobgy, A.H. Adv. Chem. Ser. Am. Chem. Soc. 1974, 133, 376-392
81. Nauman, E.B. Chem. Eng. Sci. 1975, 30, 1135-1140
82. Spielman, L.A.; Levenspiel, O. Chem. Eng. Sci. 1965, 20, 247
83. Kattan, A.; Adler, R.J. Chem. Eng. Sci. 1972, 27, 1953
84. Villermaux, J. to be published
85. Treleaven, C.R.; Tobgy, A.H. Chem. Eng. Sci. 1972, 27, 1653
86. Treleaven, C.R.; Tobgy, A.H. Chem. Eng. Sci. 1973, 28, 413-25
87. Ritchie, B.W.; Tobgy, A.H. The Chem. Eng. J. 1979, 17, 173
88. Mehta, R.V.; Tarbell, J.M. Private communication
89. Makataka, S.; Kobayashi, J. Int. Chem. Eng. 1976, 16, 148-154
90. Dudukovic, M.P. Ind. Eng. Chem. Fundam. 1977, 16, 385-388
91. Dudukovic, M.P. Chem. Eng. Sci. 1977, 32, 985-994
92. Dudukovic, M.M. AIChE Journal, 1977, 23, 382-385
93. Takao, M.; Murakami, Y. J. Chem. Eng. Japan 1976, 9, 336-338
94. Takao, M.; Nomoto, O.; Murakami, Y.; Sato, Y. J. Chem. Eng. Japan 1979, 12, 408-410
95. Parini, K.; Harris, T.R. The Canad. J. Chem. Eng. 1975, 53, 175-183
96. Wood, T.; Hop, N.H. Proceed. Second Europ. Conf. Mixing, Cambridge, 1977, BHRA Fluid Engineering
97. Spencer, J.L.; Lunt, R.R. Ind. Eng. Chem. Fund. 1980, 19, 142
98. Beek Jr, J.; Miller, R.S. Chem. Eng. Prog. Symp. Series 1959 55, 23-28
99. Ottino, J.M.; Ranz, W.E.; Macosko, C.W. Chem. Eng. Sci. 1979 34, 877-890
100. Ottino, J.M. Chem. Eng. Sci. 1980, 35, 1377-1391
101. Ranz, W.E. AIChE Journal, 1979, 25, 41-47
102. Ottino, J.M. AIChE Journal 1981, 27, 184-192
103. Ottino, J.M.; Ranz, W.E.; Macosko, C.W. AIChE Journal, 1981
104. Ranz, W.E. AIChE Journal 1982, 28, 91-96
105. Ottino, J.M. J. Fluid Mech. 1982, 114, 83-103
106. Chella, R.; Ottino, J.M. To be presented at ISCRE 7
107. Ottino, J.M.; Macosko, C.W. Chem. Eng. Sci. 1980, 35, 1454-57
108. Klein, J.P. David, R.; Villermaux, J. Ind. Eng. Chem. Fundam. 1980, 19, 373-379
109. Zoulalian, A.; Villermaux, J. Adv. Chem. Series. Chem. React. Eng. II Evanston, 1974, 133, 348-361
110. Villermaux, J. to be published
111. Belevi, H.; Bourne, J.R.; Rys, P. Helv. Chem. Acta, 1981, 64, 1618-1629

112. Belevi, H.; Bourne, J.R.; Rys, P. Chem. Eng. Sci. 1981, 36, 1649-1654
113. Bourne, J.R.; Kozicki, F.; Moergeli, U.; Rys, P. Chem. Eng. Sci. 1981, 36, 1655-1663
114. Angst, N.; Bourne, J.R.; Sharma, R.N. Chem. Eng. Sci. 1982 585-590
115. Bourne, J.R.; Rohani, S. Proceedings of the Fourth Europ. Conf. on Mixing, April 27-29 1982, Paper J2, 313-325
116. Bourne, J.R.; Schwartz, G.; Sharma, R.N. Proceedings of the fourth Europ. Conf. on Mixing, April 27-29 1982, (BHRA Fluid Engineering), Paper J3, 327-338
117. Pohorecki, R.; Baldyga, J. Third Internat. Summer School. Modelling of heat and mass transfer processes and chemical reactors, Varna, May 21-30 1979.
118. Miyawaki, O.; Tsujikawa, H.; Uraguchi, Y. J. Chem. Eng. Japan 1975, 8, 63-68
119. Rys, P. Angew. Chem. Int. Ed. Engl. 1977, 16, 807-817
120. Nabhotz, F.; Ott, R.J.; Rys, P. Helv. Chim. Acta 1977, 60, 2926
121. Belevi, H.; Bourne, J.R.; Rys, P.; Helv. Chem. Acta, 1981, 64, 1630-1644
122. Bourne, J.R. AIChE 1981 Annual Meeting, New Orleans, Nov 8-12 Paper n° 61a
123. Spalding, D.B. Levich Birthday Conf. on phys. chem. hydrodynamics, Oxford, 11-13 July 1977
124. Bourne, J.R.; Rys, P.; Suter, K. Chem. Eng. Sci. 1977, 32, 711
125. Nabholtz, F.; Rys, P. Helv. Chem. Acta 1977, 60, 2937-43
126. Bourne, J.R.; Kozicki, F. Chem. Eng. Sci. 1977, 32, 1538-39
127. Belevi, H.; Bourne, J.R.; Rys, P. Helv. Chem. Acta 1981, 64, 1599-1617
128. Barthole, J.P.; David, R.; Villermaux, J. Proceedings ISCRE 7 Boston 1982, Oct. 4-6
129. Bourne, J.R.; Kozicki, F.; Rys, P. Chem. Eng. Sci. 1981, 36, 1643-1648
130. Takao, M.; Yamato, T.; Murakami, Y.; Sato, Y. J. Chem. Eng. Japan, 1978, 11, 481-486
131. Murakami, Y.; Takao, M.; Nomoto, O.; Nakayama, K. J. Chem. Eng. Japan 1981, 14, 196-200
132. Hanley, T.R.; Call M.L. 74th Annual AIChE Meeting, New Orleans Nov 8-12, 1981, paper 18d
133. Ghodsizadeh, Y.; Adler, R.J. 74th Annual AIChE Meeting, New Orleans, Nov. 8-12, 1981 paper 61i.
134. Bhatt, B.L.; Ziegler, E.N. AIChE Journal, 1977, 23, 217-224
135. Bryant, J. Adv. in Biochem. Eng. 1977, 5, 101-123
136. Costa, P.; Lodi, G. The Canad. J. Chem. Eng. 1977, 55, 477-79
137. Bourne, J.R.; Crivelli, E.; Rys, P. Helv. Chem. Acta, 1977, 60, 2944-2957
138. Sasakura, T.; Kato, Y.; Yamamuro, S.; Ohi, N. Internat. Chem. Eng. 1980, 20, 251-258
139. Rachez, D.; David, R.; Villermaux, J. Entropie, 1981, N° 101, 72, 32-39
140. Bryant, J.; Sadeghzadeh, S. Proceed. Third Europ. Conf. on Mixing, April 4-6th 1979, York, BHRA Fluid Eng. paper F3, 325
141. Bryant, J.; Sadeghzadeh, S. Proceed. Fourth Europ. Conf. on Mixing, April 27-29 1982, BHRA Fluid Eng., Paper B4, 49-56
142. Brennan, D.J.; Lehrer, I.H. Trans. I. Chem. Eng. 1976, 54, 139
143. Khang, S.J., Levenspiel, O. Chem. Eng. Sci. 1976, 31, 569-77

144. Okamoto, Y.; Nishikawa, M.; Hashimoto, K. Int. Chem. Eng. 1981, 21, 88-94
145. Van der Molen, K.; Van Maanen, H.R.E. Chem. Eng. Sci. 1978, 33, 1161-68
146. Nishikawa, M.; Okamoto, Y.; Hashimoto, K.; Nagata, S. J. Chem. Eng. Japan 1976, 6, 489-494
147. Fort, I.; Placek, J.; Kartky, J.; Durdil, J.; Drbohlav, J. Collect. Czechoslov. Chem. Comm. 1974, 39, 1810-1822
148. Mann, R.; Mavros, P.P.; Middleton, J.C. Trans. I. Chem. Eng. 1981, 59, 271-278
149. Mann, R.; Mavros, P. Proceedings of the Fourth Europ. Conf. on Mixing, April 27-29, 1982, BHRA Fluid Eng. Paper B3, 35-47
150. Joshi, J.B.; Pandit, A.B.; Sharma, M.M. Chem. Eng. Sci. 1982, 37, 813-844
151. Rod, V.; Misek, T. Trans. Inst. Chem. Eng. 1982, 60, 48-53
152. Oldshue, J.Y. A.C.S. Symp. Series, 1977, 55, 224-241
153. Oldshue, J.Y. Chemtech. September 1981, 554-561
154. Kipke, K. Chem. Ing. Tech. 1982, 54, 416-425
155. Quraishi, A.Q.; Mashelkar, R.A.; Ulbrecht, J.J. AIChE Journal 1977, 23, 487-492
156. Ulbrecht, J.J.; Sema Baykara; Z.; Chem. Eng. Commun. 1981, 10 165
157. Lane A.G.C.; Rice, P. Trans. I. Chem. E. 1982, 60, 171-176
158. Okamoto, Y.; Nishikawa, N.; Hashimoto, K. Int. J. Chem. Eng. 1981, 21, 88-94
159. Pahl, M.H.; Muschelknautz, E. Int. Chem. Eng. 1982, 22, 197
160. Boss, J.; Czastkiewicz, W. Int. Chem. Eng. 1982, 22, 362-367
161. Matsumara, K.; Morishima, Y.; Masuda, K.; Ikanaga, H. Chem. Ing. Tech. 1981, 53, 51-52
162. Henzler, H.J. Chem. Ing. Tech. 1979, 51, 1-8
163. Gross-Röll, F. Int. Chem. Eng. 1980, 20, 542-549
164. Bürgi, R.; Tauscher, W.; Streiff, F. Chem. Ing. Tech. 1981, 53, 39-42
165. Nauman, E.B. The Can. J. Chem. Eng. 1982, 60, 136-140
166. Nauman, E.B. AIChE Journal, 25 (1979), 246-258
167. Murakami, Y.; Hirose, T.; Ono, S.; Eltoku, H.; Nishijima, T. Ind. Eng. Chem. Process Des. Dev. 1982, 21, 273
168. Brauer, H. Chem. Ing. Tech. 1980, 52, 992-993
169. Le Goff, P. Proceed. 2nd Europ. Conf. on Mixing, Cambridge, March 1977
170. Ries, H.B. Int. Chem. Eng. 1978, 18, 426-442
171. Müller, W. Chem. Ing. Tech. 1981, 53, 831-844
172. Sommer, K. Fortschritte der Verfahrenstechnik 1981, 18, 189
173. Nauman, E.B. J. Macromol. Sci. Revs. Macromol. Chem. 1974, C10, 75-112
174. Gerrens, H. Proceed. 4th Int. Symp. Chem. React. Eng. Heidelberg, 1976, 584-614
175. Villermaux, J. Colloquium of the Working Party C.R.E. Novare, may 25-26, 1981
176. Sahm, P. Thèse, Inst. Nat. Polytech. Lorraine, 1978
177. Olson, J. H.; Stout, L. E. In "Mixing--Theory and Practice," Vol. II, Uhl and Gray, Eds.; Academic: New York, 1967.

RECEIVED April 15, 1983

INDEX

INDEX

D

E

F

K

L

M

S

Z

Jacket design by Kathleen Schaner
Indexing and production by Florence Edwards and Paula Bérard

Elements typeset by Service Composition Co., Baltimore, MD
Printed and bound by Maple Press Co., York, PA